To Jim Squires, a great editor
and a writer first.

Jack Fuller

Convergence

Convergence

JACK FULLER

DOUBLEDAY & COMPANY, INC.
GARDEN CITY, NEW YORK
1982

Library of Congress Cataloging in Publication Data

Fuller, Jack.
    Convergence.

    I. Title.
PS3556.U44C6   813'.54
AACR2
ISBN: 0-385-18023-3
Library of Congress Catalog Card Number: 81-43483

For Alyce

There was no certainty; only the appeal
to that mocking oracle they called History,
who gave her sentence only when the jaws
of the appealer had long since fallen to dust.

Arthur Koestler

ONE
Washington, 1978

1

Across the river the soldiers' stones shone in amber ranks upon the cemetery hill. The Washington Monument's long shadow on the mall gave its last reading of the hour and then faded into shadows. A few runners still pounded away on the grass. Lights blinked out in the buildings that ringed the green. Flags limped down poles into unceremonious arms.

It was finally time for all the powerful and pretenders to lock away their documents, to return the calls they had put off until they could be sure there would be no answer, to surrender their offices to the permanent janitorial staff whose job it was to maintain them from occupant to forgotten occupant. It was finally night, and the temporary stewards of government had proven once again not only that they were manly enough to take the grinding pace of national destiny but also that they were essential enough that this was required of them. Those without a future packed up their attaché cases and drove home to wife and family. The ambitious had further duties. A late drink at an embassy. A fashionable entrance between acts at the Kennedy Center. Dinner with friends and enemies. Or if there were nothing else, at least a short appearance among the correspondents at the Class Reunion, where a secret could be quietly shared, a reputation discreetly tarnished.

Richard Harper's shift was just beginning as the officious aides and their principals made their way from desks to other assignations. He was already tired, eyes red with the polluted air, and his business was to bear it and to wait.

For weeks Washington had gone without a breeze, the atmosphere piling up with fumes and ozone, the gray ambiguous haze hiding the sky. Now until September every day would be the same: cloudy or fair, you could not tell the difference. They said it was auto exhaust and the swampy location, but Harper knew better. The obscuring haze was industrial pollution, a noxious mixture of wasted breath and oxidized hopes. It turned to poison in the sun. The industry, of course, was government.

Harper worked for the government himself, but he was not a public man. In his job a person hid his work habits, was vague in conversation about his responsibilities. His strength was in concealment, his pride in obscurity. If he knew something, he did not let on that he did. And nearly everyone assumed that he knew even more. Harper worked for the Central Intelligence Agency.

It was as hot as Quito or KL or any of the CIA's other tropical listening posts, but Washington was not the Agency's turf. It belonged to the FBI, first by gentlemen's agreement and later, when no one believed anyone was a gentleman anymore, by law. The Agency joined the game within the United States now only by express invitation, and then only to observe and consult. Harper was along on this operation as part of his orientation course for a new assignment. He had been transferred from the analysis side to internal inspections. When he had inquired into the reasons, he had been told to consider it an ingrade promotion. The Director of Central Intelligence was strong on inspections, Harper was told, and he wanted to expose the men of greatest promise to its friendless discipline. In the old days, the inspections unit was where they put scared old men who could be controlled. But now the word was out that the DCI had made a change. If you wanted to get ahead in the Agency, you had to put in your time as a snitch. And spending an evening on routine physical surveillance duty with the Sisters, as the FBI was called at Langley, was part of the training.

We started calling the FBI the Sisters when the Bureau began to boast of its female recruitment figures. The name even crept into some of the speeches given by the DCI, who liked to refer to the Bureau as "our sister agency." The smile this tickled at Langley was just another of our secrets. In the Agency, you see, there was what could be called a healthy skepticism about the quality of the work done by the intelligence personnel of the Bureau's Division Five. When something came off badly, it always seemed traceable to their lack of finesse. Of course, this was quite convenient; one agency or the other always had to take the blame. And so every new inspector was expected to spend an evening with the Sisters as they doggedly trailed KGB agents around town. This way he could learn the Sisters' procedures and weaknesses firsthand. Within the Agency, this had come to be known as "covering our interface."

Malloy, the driver, pulled into a space along the park that lay between the Hay-Adams Hotel and the White House. In front of them was a Mercedes; behind, a Volvo. Their unmarked beige Ford was about as inconspicuous as a dray horse because in Washington nobody bought American. The old streetlights, preserved since the conversion from gas, shed more charm than light. But that was all right. The agents had starlight field glasses, and from their vantage they could clearly see the H Street sidewalk, alongside the hotel, and the alley in the rear. They could not see the entrance. It faced away from them onto Sixteenth Street. But the other car, which had covered the mark on his trip to the hotel, watched it from a spot across the park.

"Might as well crank open those windows, men," said Malloy as he turned off the engine. "It's going to be awhile."

The choice was either to keep the air conditioner on and

overheat the engine or to turn it off and overheat the men. Bureau regulations, said Malloy, were clear. "It's destruction of government property either way. Cars are harder to replace."

Malloy was a large man, long across the shoulders, thick-necked, and endearing in the childlike way some strong men have when they seek the approval of the weak. His partner, Torchiana, was small, dark, and wiry. He made it quite apparent that he sought approval from no one. Both were about the same age, a good fifteen years younger than Harper, and both had come to the Sisters via the Marines.

At Langley it was a standing joke how many of the Sisters had been in the Corps. The Bureau even had its training facility at Quantico, and recruitment was often nothing more than a change of uniform. If the Agency was Ivy League, the Bureau was Fordham night. This was a source of endless delight, of course. As one of our Founding Fathers put it during the time when the rivalry was most intense, "The Marines and Mr. Hoover's Bureau may have different missions, but they surely share the same caste."

"When you signed on," said Harper, "did you figure this was the way you'd be spending your evenings?"

"Volunteered for it," said Torchiana, sliding lower in the front seat. "Plenty of opportunities. Division Five's been light at the top since the purges."

The purges had virtually run their course by the time Malloy and Torchiana had joined the Sisters, and the Left had disappeared into the Center like a cutpurse into a crowd. To these two, the purges were not real. They were a newspaper clipping, an argument over beer after class. These men were of the new era, and they probably thought everything had changed.

Harper had been posted in Africa during the purges. He had heard all the corrosive rumors from Washington, tales of disgrace and ruined careers. But he had not been touched personally. During the long night of suspicion—not of treason this time but rather of unwarranted zeal—Harper's conduct had never raised a doubt. He had a reputation as a methodical, careful agent. He was reliable, not in the sense of loyalty, which now seems to have become passé, but rather in his steady conformity with expectations. He had had his moment of triumph in the arena, but even this did not make him suspect. During the ugliest moments of the purges, no one ever questioned the legitimacy of that operation—the one we called Black Body.

When the analysts announced their conclusion that the Soviets had deployed the new Black Body communications security system, the atmosphere at Langley turned as grim as war. We had been called together for a rare meeting of the entire top staff behind the heavy steel doors of the secure conference room in the DCI's suite. Harper did not learn of the meeting until much later, of course. It was an affair for principals only. Most of us had no idea what had come up to require such a gathering, but we knew it was trouble when the DCI began by apologizing for failing to brief us on paper in advance, saying that the matter was too sensitive to commit to writing. Only very bad news was this closely held.

After the preliminaries the DCI turned the meeting over to Danzig, head of the technical research staff. He was a small, foppish man, and he stood with his fingertips warming in the pockets of his vest as he spoke.

"Gentlemen," he said, "as you know, for some time now counterintelligence has been reporting a decline in observed activity in the Soviet deep-cover networks in all theaters. Until recently this was unexplained. I know there has been some speculation that perhaps a shake-up in the KGB's hierarchy at Lubyanka had put their illegal operations in a temporary holding pattern. Unfortunately, this has proven not to be the case."

There was a discernible rustle of activity around the table. Simpson, who had floated the silly idea about instability at Lubyanka in the first place, leaned forward and put his elbows on either side of his coffee cup, his hands clenched before his face in fists. Myers cleared his throat again and again. He had lent his name to Simpson's foolish theory. The others, who had not declared themselves, shifted in their seats. A new grouping was forming, an alliance against Simpson and Myers. I did not have to change my position. I operate on the principle that with the Soviets there is always ample reason for assuming the worst.

"Coincidental with the reduced observations of deep-cover activity," said Danzig, "the National Security Agency began picking up some curious emanations from Soviet installations in various locations. At first it was believed that these might relate to their efforts to monitor our nets, but it has since become clear that the signals are part of a new system by which the Soviets can communicate directly and without human contact with their deep-cover operatives. And, gentlemen, I must tell you that we have no foreseeable method of compromising these messages."

"Impossible," said Simpson. "Strong codes are too cumbersome."

"This is not precisely a cipher technique," said Danzig. "Would that it were. This seems to be an electronic method that is its own security mechanism. And, to be frank, at this point we have no idea how it works. We do not know how much information can be passed in each brief transmission. There is no effective means of triangulation. We do not know what kind of device is used to receive or transmit, although we fear that it is devilishly small. We are devoting maximum resources to the problem, of course, and in time we may develop some better idea of its capabilities and weaknesses. But I need not tell you how vulnerable we will be in the meantime. All I can say is that this interim period might be very long.

"The problem as I see it," Danzig went on, "is to come up with some nontechnical countermeasure to buy our scientists time. I do not profess to know what form this might take. Tricks of nature are my area, tricks of malice yours. But I can say that the priority is the highest."

If the Soviets could talk to their deep-cover illegals without physical contact, then we had no window at all on their operations. Not only had we no way of knowing what information was being passed, but we also had no handle on who the illegals were. Black Body eliminated the need for human intermediaries. We could follow every embassy second secretary and chauffeur, every student and visiting artist, for twenty-four hours a day and still not get a line on the expanding network of deep-cover spies. Black Body was putting us completely in the dark.

"It appears," said the DCI, "that the new system is only being used by new assets. None of our old standbys have been detected sending the signals. Moving in on them will get us nowhere. But others are penetrating us, and we don't know where. The damned thing is tight, gentlemen, very tight."

"But just how sure are we?" said Simpson. "Aren't we in danger here of overreacting?"

"I don't think there is much risk of that in this case, George," said the DCI, and there was an appreciative titter around the table at the rebuke. "More discussion at this point would not be particularly fruitful. We know the problem. We will convene again tomorrow to discuss our options."

In the end we were driven to the human option. The burden fell to Harper. And by all objective measurements he carried the mission off. He mounted a deception against one of the most clever KGB agents in the field, Anton Ignatyeff Kerzhentseff. That led to Harper's pick of outside assignments, his eventual return to Langley, and his tour of duty in the inspections unit. Though stakeout duty at the Hay-Adams with the Sisters might not immediately suggest it, he was on the way to the upper ranks. And this marked him out to me as someone who was certainly a man to watch.

Harper loosened his durable, blue-striped tie and gently rolled his collar button free. The shirts were one of his steel habits. For more than twenty years he had bought them all from the same shop. They had the right oxford cloth, the right geometry of cut and style, the right collars, buttoned down. He picked them up by the dozen whenever he chanced to be in New Haven. Otherwise, he ordered them by mail from all his far postings. It pleased him that the store was always prompt and understated, always sure of the best method of delivery, even to the remotest places. Never a fuss. He liked to be taken for granted. As for suits, there were seven in summer, seven in winter. Harper was a perfect Brooks 40, so he bought off the rack, two a year. Each suit remained in service for seven years and then was replaced by another of similar color and cloth. The discards went to Goodwill or, overseas, to his servants. There were those in the Agency who thought Harper might loosen up once he came inside. Some men, when they are operational, exercise Prussian self-discipline over the small things because the large ones are always verging out of control. But Harper did not ease up when he returned to Langley. Balmy Tuesdays were seersucker, and that was that.

Harper did not feel much like an inside man as the evening dragged on in sweat and confinement. The back seat of the Ford was too shallow for his legs, and it angled imperfectly to the cant of his back. He turned and leaned into the corner between the armrest and the cushion, stretching his legs awkwardly over the transmission hump. His arm crooked across the back of the seat, just a little too high. In time, his hand fell asleep.

"Gets a little tedious, doesn't it?" he said, meaning no offense.

"Gets hairy sometimes," said Malloy.

Torchiana was for some reason irritated. His face, a sore only recently healed, turned toward Harper.

"War of nerves," he said.

"I guess it's the only war we've got," said Harper, leaving it at that.

Across the park, directly opposite the hotel, the White House was lighted up like a stage set. The only sign of life was at the gate where the guards now and then let a visitor in, a limousine out.

"They could keep a pretty good eye on Ivan from the living quarters," said Harper, raising his numbed hand toward the glowing windows behind them.

"And Ivan could keep a pretty good watch on them," said Torchiana. It was always amusing to see a man who believed in his work.

"This party at the Hay-Adams," said Malloy over his shoulder, "it's some kind of trade affair. Our guy's in there on the legit. It's his cover. He'll stumble out sooner or later and go home."

"Which is where?"

"One of their apartments," said Malloy, "the one where the idiot threw the bundle of documents that time."

"*Our* idiot, you mean," said Harper.

"I didn't want to get personal."

Harper nodded. The idiot had been one of the Agency's men who was destroyed in the early stages of the purge. He had carried with him copies of everything in his safe, mostly garbage but all of it code-word classified nonetheless. It had taken him about a year on pension before he had finally cracked. He had wrapped up the junk in plain brown paper and sauntered over to one of the Soviet residences. Pitched the whole bundle over the fence and kept on walking, just like that. The other side didn't know what to make of it. They really don't understand us very well, think we're damned erratic, capable of anything. Somebody got scared that the package might be a bomb. So they called the Sisters, of all things, and a couple of demolition boys retrieved it. When the package turned out to be nothing but paper, the word leaked out immediately. The stories in the newspapers embarrassed the KGB as much as they did the Agency. Two agricultural advisers were on an Aeroflot flight home before the week was out. It ruined three good men at Langley too. And, though the news accounts never mentioned it (a clue to their source), a couple of people at the Hoover Building also took a fall because until the KGB made the phone call for help, nobody had even seen the package being tossed.

"The Bureau did vigilant work that night, all right," said Harper, putting it back to them.

"Maybe we saw more than we said we did," said Malloy. "It doesn't pay to boast."

Harper rubbed feeling back into his tingling hand and acknowledged the point. Malloy was young, but he knew where to hide.

"When the mark gets home," Malloy went on, "we turn him

over to the passive unit and call it a night, except for the paperwork, of course. Torchiana takes care of that. He's junior. I'm only sorry we aren't providing you a better show tonight."

"It's very instructive," said Harper, trying not to shade the words. Torchiana picked up on them anyway.

"What do you mean by that?" he said.

"They're careful. You're careful," said Harper, being careful too. "You have to take what comes by. Read the tea leaves. That sort of thing. It's the same in our shop."

"You sound like a Brit. You know that?" said Torchiana. "Just like a Brit."

Harper checked his watch and was dismayed at how little time had passed. Hardly enough for the mark, whoever he was, to have finished his first polite drink. A skinny dog meandered down the sidewalk, mildly curious about signposts and the stumps of trees. When it reached the car it paused and looked up at Harper with tired, rheumy eyes. Then, finding nothing of particular interest, it moved on. Harper whistled to it softly, hoping to get it to bark. Anything to break the silence. It paid no attention. Up in front the two of them took turns dozing. He could hear himself breathe.

Harper had a low tolerance for silence. At home and in his office he played the radio constantly, and it was often tuned to talk. In Japan it was Asian chatter. Those who did not know him thought he was trying to learn the language. Those who did thought he was trying to frustrate any unauthorized ears. But neither was the case. His duty was to be taciturn, but he craved to hear the sound of other voices, chatter, song. If you asked him about this, he would laugh it off, saying he needed to drown out the whispers of his conscience. That was a good enough answer; at the Agency, we have to tolerate some eccentricity—even, grudgingly, a certain amount of remorse.

"Maybe you could tell me who it is we're waiting for," Harper said finally. Torchiana waved the question away and lay back against the seat, but Malloy rose to it.

"They didn't brief you?" he asked.

"They briefed me," said Harper. "I got the organization

chart on the viewgraph, the slides of master spies you have busted over the years—a pretty unimposing lot from their pictures. I even got the little lecture on the latest reforms. Just like the visiting sheriff."

"Maybe your ticket's not punched," said Torchiana, yawning.

"I've got all the clearances," said Harper. "I've got clearances it's a crime to say their names out loud. My ticket is well punched, thank you."

"Hey," said Torchiana, "no need to get pissed."

"We're talking top secret," said Malloy.

"Rest assured," said Harper.

"He's a KGB colonel under diplomatic cover," said Malloy. "He's been around. But the thing is, he doesn't seem to have made a move here yet. We've kept an eye on him, but it's been the party circuit and that's about all. It just doesn't fit. This guy's career has been a hundred percent mischief. And the one thing we know for sure about these types: they never go straight."

"He drives his own car, a Chrysler, black," said Torchiana.

"They love the big boats," said Malloy, turning to Harper. "The vehicular side of basic Marxism-Leninism."

It was just like a Russian to indulge our vices grandly. In so many ways, Harper thought, they had already become just like us—only more so.

"Did I mention the guy's name?" said Malloy. "Kerzhentseff. Anton Ignatyeff Kerzhentseff. Heard of him?"

The name exploded on Harper. Kerzhentseff was, of all the thousands of KGB operatives in the world, one of the most virulent. He was the man Harper had gone up against in Black Body. Harper knew Kerzhentseff, all right, knew him the way you can only know and fear someone you have deceived.

Malloy was watching him in the rearview mirror. Harper met the eyes and kept his face steady as a fist.

"You pronounce the name very well," Harper said.

"My mother was Polish," said Malloy. "She hated the bastards."

No one in intelligence work can afford to believe too deeply in coincidence, especially when it involved someone like Ker-zhentseff. Harper's instincts told him that somehow there must have been a mind behind this meeting. In espionage, chance is always a man.

Harper had known this feeling before. Everyone who has spent time in the field has suffered it. It can burst upon you with the clarity of a muzzle flash in the night or sneak up like the twilight. But however it comes, it changes everything. The betrayer wonders to whom he has been betrayed. Your cover, worn like a second skin, suddenly splits. And you don't know who in the world you are supposed to be.

Harper did what any trained agent does when the wound tears open: he went on as though nothing had changed. His best guess was that the two others in the car knew nothing of his previous connection with the Russian mark. The Sisters did not play their shots on the carom. The Agency might like to make a Judas kiss off three cushions, but the Sisters preferred to hit straight and hard. This he felt quite sure of, but not quite sure enough to act on.

"What kind of paper do you file when this is over?" Harper asked.

"Ins and outs," said Torchiana. "Times, locations, routes. They want all the details. Our guy bums a match from some mope on the street, they want a name and Social Security number. If Ivan pops in somewhere to take a leak, somebody has to go in and check it out after."

"I suppose it all adds up to something," said Harper.

"Beats me," said Malloy. "Isn't the drill the same when you guys do it?"

Harper was on his guard not to seem defensive.

"We're short on manpower overseas," he said. "We can't follow everybody. We concentrate on targets of opportunity."

"Hey," said Torchiana. "Were you a grunt?"

"Pardon?"

"Targets of opportunity," he said. "For a minute there you sounded like a regular soldier."

"It's just a phrase," said Harper.

"I thought maybe OSS or something."

"That was before my time," said Harper. "I did a tour in Vietnam."

"No kidding," said Torchiana. "When?"

"Late sixties, very late."

"We were there in seventy. Fifth Marines. I was in a rifle platoon. Malloy here had it easy. No humping the bad bush. Targets of opportunity."

"Roger that," said Malloy.

"Shitty little war," said Torchiana.

"Where were you stationed?" asked Malloy.

"Saigon mainly."

"That wouldn't have been hard to take," said Torchiana. "I went there once. Had to take shots when I got back."

"It all depends on your frame of reference," said Harper.

"Ever get into the deep shit?" asked Torchiana.

"Deep enough," said Harper.

"Firefight?"

"They were doing all the firing."

"You must have been out with the ARVN."

"It was different for the spooks, Tony," said Malloy.

"Is that right, Harper?" Torchiana taunted. "Was it very different for you?"

"It wasn't a walk in the country," said Harper.

He did not want to go into it any further just then. Even under the best of circumstances, it was a subject he avoided. And this was not the best. Close, too close, the queasy sense that something malign and invisible has come into play. If a wound was pulling open now, it was a wound he had first suffered there.

But he had healed, hadn't he? The fever dreams had subsided. He had never given way to them, not completely. Sometimes he thought he was one of the few who hadn't. The Agency was full of men Harper thought had lost their moorings to anything at all that is as it seems.

Of course, he accepted a healthy skepticism, even nurtured it in himself. It was part of the job. But that did not mean you had to worship the Mocking God—the God of the lie, of disinformation, false appearances, and doubt. This was the cult of possibility. And it was a disease. Its followers at Langley were willing to imagine anything: That the latest Soviet defector was a plant. That those who said he was genuine knew otherwise. That Langley itself was controlled at its highest levels by demons from the other side. Even that the rift between the Soviet Union and China was an elaborate geopolitical ruse to lull the West into a false sense of security. Yes, it was possible, just possible—once the idea of disinformation came into it—as possible as any nightmare.

For Harper, possibility was the Mocking God. And no philosopher of doubt has ever lived its creed so completely as the cultists at Langley did. For the philosopher, it was a matter of pure speculation. But for the cultists, it was a matter of cold, mortal practice.

Harper liked to believe that for his own part he had developed an immunity against this disease. His wound had become infected by it in Vietnam, but he had fought off the poisons and survived. And now there was nothing metaphysical about him. He was a positivist of espionage—like the Sisters, whose common sense he openly admired—a dull, plodding positivist. He held close to what he could see and touch, to the data. And that evening at the Hay-Adams, even as the Mocking God whispered to him like conscience in the silence, he kept his mind on the certainties: the heat, the mission, and Kerzhentseff.

Some sort of affair was letting out at one of the swank old homes along the park. They were too expensive for anyone to live in anymore, so they had all been converted to causes, charitable and otherwise. Harper watched the people idle at the door. The men wore white. The rich and their waiters always dressed alike.

"It's some kind of conference on poverty," Malloy explained.

"Looks like it did some good," said Torchiana.

A few couples wandered in the direction of the darkened car, and some of them glanced curiously at the three men hunched up in it. Torchiana kept his field glasses down between his knees and smiled sweetly as the couples passed.

"Diplomats," he spat when they had all found their cars and cabs and been driven away.

"You getting stiff back there?" asked Malloy.

"Can't tell," said Harper. "Too numb."

"Not quite in shape for street work anymore," said Torchiana.

"Seat work you mean," said Harper, trying to straighten his legs.

"Look," said Malloy, "why don't you get out and stretch a little? Our guy isn't going to be coming out for a while. Just don't stray too far from the car in case we have to crank."

"And don't play in the street," said Torchiana.

Harper unlatched his door and climbed awkwardly out. With his first step he nearly fell. His foot was dead and heavy. It wouldn't bear weight. He limped back and forth using the car roof as a crutch until he got the life back into his funny foot. Then he moved out under the trees. The slightest breeze riffled through the branches and made the shadows move. He was clammy in his suitcoat. Here and there a bum lay asleep on the benches. A couple stood embracing on the asphalt path.

Harper gave them plenty of room as he moved toward the hotel side of the park. He stopped short of the sidewalk and waited there. For what? For something to burst. For a car to sweep up, haul him in, speed him away. For the click of a safety snapping off in the dark before the shot. Circumstance closed in on him. He was an outside man again, as he had been in Tokyo and Vietnam, and he was alone.

Harper was always candid about his fear. He admitted that he did not feel particularly well cut for outside work. He was brighter than he was courageous, and though that did not count for as little at Langley as it did among the Sisters, there were still dues of danger that had to be paid. But Harper felt that he had paid them in full already, owed nothing more. He had paid both the physical and the emotional parts, principal and interest, both Vietnam and Tokyo. He felt that he had earned the right to come inside now for good, to make his way from desk to desk, which was a kind of advancement. But the night was tropical and still. The traffic whispered in the street.

All the voices spoke to him. And the menace was like a memory: both memories at once, elaborate, interwoven, the violence of one and the sullen foreboding of the other.

These aftershocks are normal, they had told him when he went to the Agency doctors about his Vietnam nightmares. You might be surprised at how many are affected. No one expects you to be a machine. These things work themselves out in the darkness. How is your appetite? He did not care for Asian food, thank you. They laughed and said it proved he had not lost his critical faculties.

Of course it was normal, just as it is normal that when a man is shot he bleeds. You might be surprised at how many are affected. Harper was beyond surprise.

But not beyond the fear, not beyond the feeling of the night closing in on him, not beyond seeing again the crude, mutilated bodies in Saigon or Kerzhentseff's drawn, hollowed face and wondering what in the world these things were supposed to mean. They had reassured him after Black Body that every indication was that the mission had been very meaningful, indeed. And if the picture became less clear over time, one must expect ambiguities to creep in, visions and revisions. Nothing Harper should worry about. "We aim only for interval," they had told him. "One interval and then another. Never eternities. Eternities don't last."

At first they had humored him, letting him in on information which, strictly speaking, he had no need to know. They thought he deserved to savor his success. They rather enjoyed sharing the great joke on the wily Kerzhentseff. But the joke grew stale. The savor went out of it. And finally they cut Harper off—not meanly, of course, but with their characteristic surface grace.

He pulled his handkerchief from his pocket and blotted the moisture from his face. He was sweating as if he had run a race. He tried to concentrate on the stately old hotel across the street, gray, ornate, its high windows incandescent. It calmed him. Like the Republic, it was built grandly when that still seemed appropriate. Yet it was still only faintly weathered. Little as he cared for Washington and its public ways, this night he appreciated how accustomed he had become to feeling safe here. It was strange the way you spent your working days worrying over all the crude, amoral forces that threatened it; yet in the end it seemed that you were the one protected.

Then he saw Birch. All the protective assumptions fell away. Jerry Birch, his secret accomplice against Kerzhentseff in Tokyo, violating all the rules of probability and trust, emerged from the alley behind the hotel, paused beneath a streetlight, then strode away toward Seventeenth Street. There was no mistaking his boyish face, the open, eager, deceptive expression, the soldier's straight back and marching gait. He was right there in front of Harper as he had been so often before, observed from a distance in the night, surveilled and perhaps even suspecting he was, but not showing the slightest self-consciousness. He rounded the corner and was gone.

Harper let him go away unchallenged. And as Birch left, Harper asked after him in silence—the voice inside whispering the words—how it was that all three of them had suddenly appeared on the same block of the same city of the same protective and protected country thousands of miles from where they had done their first dark work. He asked which of them this time was the victim. The retreating figure gave no answer.

Harper waited, backed up against a tree, counting off an interminable five minutes before allowing himself to move. Then he returned to the car. Torchiana was scanning the alley through the starlight binoculars.

"Have a good walk?" Malloy asked. "You missed the excitement."

"I did?"

"Guy was in the alley. You see him?" said Torchiana.

"Maybe he was lost," said Harper.

"Malloy shot him anyway," said Torchiana. "Standard procedure. Picked him up just outside the back door of the hotel there. See it? He was leaning up against a downspout. Then he came into the light. Here, take the glasses and have a look."

Harper got into the car and lifted the heavy device to his eyes, resting his forearms on the front seat between the two men to steady his hands. It was a murky, green, underwater picture he saw, a ghost on a radar screen. But he could see the alley door and the downspout, all stark and contrasting, with no middle shades. Then the door suddenly opened and burned up the image in an explosion of light.

"What's that?" said Torchiana.

"Here," said Harper, handing the glasses back. But not before he had adjusted them down and had a look at who was standing in the doorway. It was Kerzhentseff.

"Maybe we got something tonight after all," said Torchiana. "The sonofabitch is in the alley."

"I'll call a backup," said Malloy.

"He's smoking a cigarette," said Torchiana. "You don't have to go into the alley to smoke a cigarette."

Gaulois, Harper thought. It was one of Kerzhentseff's affectations. He lighted them with a flick of the match in the left hand, the good hand. The other hung limp and useless at his side. But he did not need it. He had a way of making even his handicaps seem a kind of mastery.

"Twenty-three buckshot," Malloy said into his radio handset and at the same time fired up the ignition. "Twenty-three buckshot."

"Holy Christ," said Torchiana.

"We're ready to roll," said Malloy. "You say when."

"Shoot him, Malloy. Forget the car. Shoot him now."

Malloy brought the camera up and began flicking off frames.

"He's pretty goddamned interested in that downspout," said Torchiana.

Harper strained to see into the darkened alley. But without the glasses or the special camera he was blind.

"He's got something," said Torchiana. "Now it's in his pocket."

The door opened and Harper saw Kerzhentseff's slim shadow slip inside.

"He's gone," said Torchiana.

"You suppose it was a drop?" said Malloy. "You suppose that baby-faced dude left him something?"

"Never know," said Torchiana. "We'll want to do the downspout once the mark is gone. You keep the camera and stay on him when he leaves the hotel. I'll hop in the backup and have a closer look."

With that, Torchiana got out of the car and jogged across the park to where the other colorless car sat at the curb. Harper climbed into the front seat with Malloy.

As they waited for the Russian finally to go home, Harper rehearsed what he would say in his incident report and how he would say it. He would have to go into some detail about the background, how he had happened to be with this particular team of Sisters on this particular night. But he would have to be careful not to give the impression that he had a compulsion to explain. Purely routine duty, that was all. Imagine my surprise. That was the tone. Then he would have to be very clear that he recognized how odd it was that the incident involved both Kerzhentseff and Birch. But the language had to be right: nothing shrill. He would not let the Mocking God's voice be heard in his.

He would send his report off into the labyrinth of security channels until it got whatever attention it deserved from whoever it was who had the duty now to attend to such things. Someone would have to worry about Birch, who was running him and to what end. But that was not Harper's concern anymore, at least not in the formal sense. He had signed off that business long ago. He had earned his immunity. Let the voice haunt others.

"At least it wasn't a total bust for you," said Malloy. "We got us a little espionage."

"Is that what it was?"

"Close enough," said Malloy. "You go along for weeks, months, just sitting on your can. Then all of a sudden something like this comes out of nowhere. That's the joy of it. Anything's possible."

2

"Partial prints," said Torchiana. "I got plenty of partials off the downspout. I got all of Washington, D.C., fingering that piece of metal. You'd better hope your pictures come out nice, Malloy, because these smudges aren't going to get us anywhere."

Malloy took off his coat, revealing his shoulder holster banding him like a back brace.

"My partner the art critic," he said, draping his jacket over the back of a chair in the squad bay. "All we need is the face."

That open, boyish, utterly unrevealing face. Harper had the picture in his mind as he sat on the edge of the big gray metal desk and flipped through the rough draft report of the incident. It only mentioned his own presence in passing. And Birch was listed as an unidentified male Caucasian.

"Looks thorough to me," he said.

"I left out the part about how hot it was," said Torchiana.

"That's all you left out," said Malloy. "I'm going home. Need a lift, Harper?"

"I'm in no rush. The wife is out of town. It's probably out of your way. I live up by the cathedral."

"Come on. Let's hang it up."

"Aren't you going to wait for the film to come back?" asked Torchiana.

"Tomorrow," said Malloy. "Tomorrow is soon enough. They are always partial prints."

They took the stairs rather than wait for the slow night elevator. Their footsteps scuffed and echoed. Malloy was right, of course. All you ever get is a partial print, just enough to make you think you might know your man. He had always been sure of Birch, but now he suddenly realized that he did not know him any better than they did. Male Caucasian. A smudge.

Malloy's station wagon was parked in the lot behind the old post office building.

"Wait a minute while I clear this mess off the seat," said Malloy. He rooted around on the passenger side, tossing toys and a child's riding seat into the back. "My old lady uses this heap as a playpen. Excuse the mess. You have a family?"

"Just my wife," said Harper. He slid into the car. It stank of babies. "She's away on business."

"Liberated," said Malloy. "I wish I could liberate my old lady to go out and pick up a little extra change."

At a stoplight on Fourteenth Street a top-heavy hooker in hot pants and a T-shirt approached the car ahead of them. She leaned into its open window.

"Fill 'er up?" said Malloy. Harper did not laugh.

Another woman took a few steps in their direction but then thought better of it and returned to the curb.

"You must not look hungry enough," said Malloy. "How long you say your woman's been away?"

The door of the car ahead swung open and the hooker slid in. After a little conference with the driver, her skinny arm snaked out the window and waved the other girl over.

"There's only one guy in that car," said Harper.

"Stereo," said Malloy. "Doesn't interest you, huh?"

"I find it distasteful."

He hated the sleaziness as much as Birch had. Birch had been straight, very straight. We had almost lost him because of it. Birch was solid, simple, direct. Birch was this. Birch was that. Harper could not resolve or rid himself of all the conflicting things Birch was. Birch was simply Birch, whoever he was.

Soon they were out of the bad zone and skimming up Massachusetts Avenue. Embassy Row was deserted except for an occasional taxi. Winston Churchill stood alone in front of the British chancery, pugnacious, daring. The evening's parties were all over, the pleasant civilities shared over coffee and cognac, the latest veiled warnings lodged. Now the gates were all locked, and armed men protected small, uneasy sovereignties.

Malloy dropped Harper off at the corner of Wisconsin, only a few blocks from his place. Harper thanked him and watched the car pull away. He stood there distracted, uttered a word, speaking to himself the way he often did when there was no other voice, no ear to overhear. He should have invited Malloy over for a drink. It would have been only polite to have done so, only safe. His body stiffened with a jolt. Nerves. He needed to sort things out, regain control.

The sidewalk sloped downward, a straight shot into the shadows. There was no one out this late at night. Harper set off, letting himself pick up speed. If anyone were watching him, where would they be positioned? Not a chance. They were simply not that good, he assured himself. And we weren't either.

The heat had let up, but Harper was sweating again. He heard the guttural sound of a car approaching from behind. Again, as he had in Lafayette Park, he fought the feeling that all the perils of the world were upon him. The car rolled up beside him, gliding even. He began to trot. The car kept up.

"Say," said a voice. Harper turned sharply and faced the danger. The buildings along the walk were all locked. There was nowhere to run.

The car stopped abruptly alongside him. Two men were inside. Whose? They could have been anyone.

"Say, do you know which way it is to Georgetown?"

Harper waited a moment until the dizziness subsided and he was sure of his own voice.

"Straight ahead," he said. "A mile, mile and a half."

"Thanks, buddy," said the man, whoever he was. The car pulled away.

It was irrational for Harper to be so unnerved by the encounter at the Hay-Adams. He knew that. It was a question of grip or slippage. But he also knew that someday the Black Body scheme would be discovered, if it hadn't been already. And then there would be revenge. At its meanest and most skilled, what we do to one another in our trade is like Dryden's description of satire: "That fineness of stroke that separates the head from the body, and leaves it standing in its place." No other wound so calls out for retribution.

As he turned the corner into his street, Harper noticed that something was wrong. Most of the houses were dark, but in his own a light burned in an upstairs window. Harper stopped on the sidewalk and peered up at the brightened blind. His wife? Impossible. She could not have returned so soon. He had received her message late that afternoon. She had arrived safely on the Coast and said she already missed him. He had left the house in the morning after she was gone, and he was not in the habit of forgetting to turn things off before locking up. There was only one possible explanation for the light. Someone had been inside after him.

He watched for shadows but saw none. He stepped quietly to the door and slid his key into the dead-bolt lock. Turning it gently, he had his confirmation. There was no resistance. The bolt had already been thrown open.

He backed away from the door and stood hidden by a hedgerow. There was no question of calling the police. The Agency insisted on the right of first refusal. But he did not want to call in the security people from Langley either. There would be questions enough about his handling of the Birch-Kerzhentseff contact as it was. One more odd note, especially if it came to nothing, and they would begin to wonder about him. Grip and slippage. What had he to worry about? they would ask. Had he taken classified material to his home? We'd better have an account of all of it. Had he reason to suspect someone might hope to find something in his home? Of course not. Socks and underwear were the most confidential things he ever kept there. But surely, they would ask, he had some reason to be concerned about the simple fact of a light in his window. Surely. He did not leave it burning, that's all. He was always careful about such things. That's what he had been taught. Ah, the unreconstructed outside man. Can't quite get used to the safer life inside, eh? The old hair across the door cliché. Perhaps Harper needs a little rest.

It was unfair and inconsistent, but it was true: after lapses of care and insufficient loyalty, the most damning thing a man could be accused of, as far as the Agency was concerned, was misplaced personal fear. Doubt was acceptable, even required. But only in the abstract, only in homage to the Mocking God. Never as sheer individual terror. He clicked open the door and eased inside.

The light from the bedroom spilled down the stairs. He closed the door without a sound and waited. The house's order had been subtly altered. Books on the coffee table lay at an abnormal skew. Chairs had wandered just a little off their usual azimuths. There was a presence in the house, impalpable, but as unmistakable as an odor. Suddenly something clattered.

He dropped behind the couch. For only the second time in his career, he wished he had a gun. And also for the second time, he decided that it was not right for a man to die on his knees. When his shallow, adrenaline breath subsided, he stood up. The noise was nothing but the sound of the old air conditioner rattling and straining against the heat.

Harper slipped out of his shoes and stepped lightly across the carpet. When he reached the kitchen he found the paring knife ready to hand on the countertop. He picked it up and held it out in front of him in death's own grip as he moved to the stairs. Halfway up, a step gave way with a sickening creak. Something stirred above him. He lunged at the bedroom door.

His knife hand dropped throbbing to his side. He did not know whether to laugh or to weep. What he saw was the woman, the woman he had told no one about, lying asleep on his bed.

She moved, stretched, but did not awaken. He wanted to

embrace her, but he held back. He had to get some distance on himself. He was losing that saving interval which gave him a chance to correct against fear and sentiment and other weaknesses. As he went downstairs to lock up properly, his palm throbbed where he had strangled the knife. He put the weapon away in the sink and poured himself some whiskey in a tumbler neat, bracing one hand with the other to steady it.

Fran would have to be warned—and sternly—against this sort of thing. There had to be ground rules. She would hate it, but it was only fair; she had made it quite clear that she had private places where she did not want him to intrude. He would have to speak to her firmly. Unannounced visits to his home were out of bounds. It was not that he never wanted to be surprised by her; the element of surprise is what attracted him. But she would have to believe that he had good and sufficient reasons for insisting that she not surprise him this way. He could make it seem mysterious. She would appreciate that. She often said he seemed sinister.

She was impressed with his secret career, especially the parts of it he refused to reveal. It rather embarrassed and pleased him to hear what she imagined: documents passed on foggy street corners, murders in sleeping cars, the bark of a .45.

"I can just see it," she had said once. "You're in Vienna. Arabs everywhere."

"I'm not sure there are that many Arabs in Vienna," he had said. "They're all in London."

"Your footsteps tap along the wet cobblestones. Someone is following. You catch a glimpse of him reflected in a store window. He wears his collar up to hide an ugly scar. But you recognize him. It is your arch-enemy, Dimitri."

"I've made enemies, Fran. But never an arch-enemy. My enemies are mostly named Bob and Bill and Tom." He did not add Anton. Anton Ignatyeff Kerzhentseff. At the time he had not been so afraid.

"You turn and confront him, the blunt blue steel of a pistol aimed steadily beneath your trenchcoat at his heart. 'Get this, Dimitri,' you say. 'Your cause is doomed. There are thousands of others just like me, ready to catch the flag as I fall.' You hear something behind you. You turn. But it is nothing. And when you look back, Dimitri has vanished. All you hear is his deep, demonic laughter echoing in the deserted street."

"What happened to the Arabs, Fran?"

"They disappeared too," she said. "Probably in cahoots with Dimitri. You never let them get the secret plans, did you Richard? You always foiled them, didn't you?"

"I am not at liberty to disclose," he said. She giggled. The game delighted her.

Harper finished his drink and returned to his bedroom. She had turned over in her sleep and lay facing him as he slipped out of his sweaty shirt and hung it on the doorknob to dry. Her legs stretched out long and bare to the thigh where the cutoff jeans began. Her face, where it had pressed against the bedspread, glowed pink as a baby's.

He had met her when he was back in the States alone looking for a house. Had he been looking for her too? He would deny it. His marriage with Janet was just fine then, still was. They knew each other's worlds. Janet had no silly illusions about him. She had seen him through the worst moments. She had watched him fail when no one was there to catch the falling flag. No, Harper would insist, he was looking only for a house.

He had gone to dinner alone in a Georgetown restaurant, nothing too fancy. There was quite a crowd, though, and too few tables. The man at the door asked if he minded sharing and pointed to a corner where Fran, blond and from a distance quite striking, sat sipping from an oversized goblet of wine. He asked how long it would be until he could have a table alone. The man at the door pursed his lips and shook his head.

Up closer, Fran was not exactly beautiful. She had her fine angles, but they did not quite come together to make a whole. Any single photograph of her would have been too flattering or too stark. You had to watch her in motion to appreciate her. And that was easy enough, for she always seemed in motion. Maybe a Picasso could have gotten her right, refracting the image into a sum of parts. But Picasso liked them round and gypsy like Janet. Fran was small and sharp and sand-colored.

"So I guess I should ask you what you do," she said when he sat down and ordered a drink. "I mean, should I know you from the news, or can I relax?"

"Don't worry," he said. "You shouldn't know me."

"I'm nobody in particular either," she said. "I just knock around with the savers."

"Savers?"

"Save the whales. Save the wilderness. Save the wetlands."

"You don't seem to share their concern."

"I share their money," she said. "There are always jobs to be had. It's one of the growth industries in this town. I need the money to support my hobbies."

"Which are?"

"I like to make things. Are you married?"

Harper laughed and caught himself turning the gold band on his finger, a talisman. He nodded.

"I was married once," she said. "Married right out of college. We liked the same bands. It lasted three years until he graduated from Georgetown Law."

"He left you?"

"Last year. We parted, actually. All very friendly. We had gotten as dull and regular as a mortgage payment. He's into real estate. He says land is the only thing that lasts."

"I'm looking for a house right now," he said, to turn the conversation in a safer direction.

"Everybody's looking for something to hang onto," she said. "They don't care what it is as long as they can keep it safe. And the only thing that gives it any meaning at all is the thought of losing it. I mean, do you know what a wetland is? It's a swamp. Nothing but a swamp."

"Lord, keep my flooded basement safe from harm," he said.

"There you go," she said. "You aren't a lawyer too, are you?"

Harper shook his head and picked up the menu.

"The pasta's very good here," she said, "if you're not committed against starch."

"I have no political views."

They ordered, and Harper made a special point as they

waited for dinner to refer to his wife, what she wanted in the way of a home, how excited she was to be coming back to the States. Fran seemed amazed when he told her he had never been divorced.

As they ate and talked, Harper found himself amused, no, intrigued by her acuteness, her youth. He eventually hinted at his work, and she took it easily. This, he imagined, was rare for a person her age.

He insisted on paying the check, and she put up little resistance. But she insisted on taking him somewhere afterward for coffee. She said she knew a good place. It was in Alexandria. They drove there in his rented car. And as she directed him through the narrow streets, he realized—well before it was too late—just where she was taking him.

"Look," she said when they got inside her door, "do you think you're going to want to do anything tonight? I mean, if you are, then there's something I have to do. I like you and everything, but I'm not exactly prepared to have your child."

"That's fair" was all Harper said.

She found it quaint that he wanted the lights out, but that night, as later, she was anything but mocking in bed. She was another woman during those moments, not cynical at all. Instead, she was totally caught up in it and not so young at all but very, very wise. He wouldn't have called what he felt love any more than she would have. But that did not mean they had a different word.

Harper hung his suit neatly on a hanger in the closet and went to brush his teeth. The whiskey had worked nicely, and he was no longer in the mood to lecture her. That could wait. They could work the matter out in the morning. He leaned over and kissed her.

"Are you awake?" he asked as she moved up against his touch.

"You kidding?" she said as her eyes came open. "I crashed hours ago."

"I should have called you today."

"No problem. I was just having a nice dream."

"How did you get in here, Fran?"

"That's my secret."

"You got a key somehow, didn't you?"

"I made one back when you first bought the place. Took it from your pocket and had it duplicated. I thought it might come in handy sometime. I'm a regular romantic."

"You gave me quite a start. I didn't know what to make of the light."

"I wouldn't betray you, Richard," she said, pleased with herself.

After they had made love, she said he had seemed prepossessed. They'd do better in the morning, she said.

"Bad business?"

"In a way," he said.

"Did you kill a man? I mean, have you?"

"Have you?" he said, and she was delighted.

He did not sleep well that night, and when he finally conceded to the dawn, he knew it wasn't any better than the night before. Whatever short dreams he had dreamed had not resolved anything. He was still beset by Birch. And in those long, middling hours on the edge of sleep, he realized that he should have disclosed his relationship with Fran to the Agency.

He had at first not taken it seriously enough to consider disclosure. It was still fundamentally not serious. That was its greatest appeal. But it had gone on for six months, and as he mulled over the unpleasant possibilities in the matter of Birch, it occurred to him that he would probably be drawn to the subject of Fran whether he wanted to or not. The Agency isn't prudish. We have no problem with this kind of infidelity, so long as everyone checks out clean. We do, as a rule, prefer that the liaisons be intra-organizational. As one rake put it, "The Agency shows a marked preference for the dipping of Company pens in Company ink." But nearly anything is tolerated, so long as it is fully disclosed through proper channels.

Harper looked over at Fran, lying there with her back to him, her hair curled tightly against her neck like a pelt. He was sorry they could not remain private. It pained him that the disclosure would have to coincide with his memo about seeing Birch and Kerzhentseff. But better that way than later, if suspicions went to full term. He slid quietly out of bed and brought his clothes into the bathroom so as not to disturb her. He had dressed and nearly finished his coffee in the kitchen when she came downstairs in one of his natty robes. She stood in the doorway and stretched.

"You didn't get all dressed up for me, did you?" she said.

"I've got to go off to work, I'm afraid."

"Don't," she said. "I had a scheme all worked out for today."

She turned on the flame under the copper kettle, and the water burst quickly to a boil.

"It's a little risky for you to come here unannounced, you know," he said.

"I thought you were big on risk. 'Danger is my business.' I thought you thrived on it."

"People could be hurt very badly."

She took the kettle off the stove and poured. The hot water foamed furiously over the coffee crystals in the brown mug Janet always used.

"We must protect them, mustn't we," she said.

"Just call, that's all. I'm not the one saying don't come at all."

"All right, Richard. I'm not going into all that again. That's settled."

He had discovered her secret one lazy Saturday quite by chance. And it had led to their only falling out. He had gone to her apartment, but she was not at home. By chance, the elderly lady across the hall was just leaving.

"Are you looking for Fran Larsen?" she asked him.

"Nothing urgent," he said. "I just stopped by to say hello."

"She usually goes down to the Torpedo Factory in the morning. I bet you could find her there."

"Torpedo Factory?"

"Down by the river in Old Town. It's a crafts center now. She has a cute little corner there. Second floor. She's so talented."

When he found the place he was surprised he had never noticed it before. It was a big warehouse of a building, not at all like the quaint shops and trendy restaurants around it. Inside, it was a warren. Potters, painters, batik-makers, sculptors, bookbinders, photographers. Each had a little cubicle on the broad factory floor. Families milled about, looking at the items for sale, picking up plates and bowls, smiling at the brightly colored prints, fingering the quilts. Upstairs, Harper wandered in and out of the little studios, nodding to the men and women who worked intently at their crafts. Finally he saw Fran bent over a flame in one of the small shops. He moved up behind her and watched. She was turning a rod of glass over a Bunsen burner. When it glowed, she pulled the material into a long filament and fixed it to the piece she was working on. The piece was a ship, delicate and transparent, rigged with thin, sparkling crystal. It was intricate and beautiful, a clipper in a child's dream. On the shelves around her he saw glasswork flowers and figurines, boys with bright diamond kites, miniature animals, and striking abstracts. Fran finally sensed him and, without taking her eyes off her work, said, "If you see anything you like, just tell me and I'll give you the price."

"I like it all," he said.

"Richard," she said, turning toward him. The softened glass rod pulled and bent in the center. "Damn. That's ruined."

"I'm sorry," he said. "I didn't realize that you had a business."

She turned down the flame and placed the rod on a square asbestos mat.

"I don't think of it that way."

"They're very graceful," he said. "It is a marvelous gift."

She seemed nervous, even embarrassed. Harper stepped toward the shelves and bent at the knees to look at one of the smaller pieces at an angle where it caught the light.

"I'm busy all day," she said. "Maybe later if you're free."

"All week is free," he said. "She was called away. I could stay and watch you make something."

"It's selfish work," she said. "You have to give it all your concentration. We could have dinner."

"This piece," he said, pointing because he was afraid to pick it up. "It's wonderful. So precise. All curves and arches. It's a kind of art, isn't it?"

He had thought to compliment her, but she was not pleased.

"It's just something I do."

"Something that lasts," he said.

"They are fragile. That's what attracts me to them. The slightest thing will shatter them. And in time the crystal I use grows brittle. A change in the temperature is enough to cause it to develop tiny cracks. Gravity alone will eventually destroy them. They are doomed to be temporary. They will not keep."

"Planned obsolescence," he said, trying to take it away from seriousness. "It keeps them coming back for more. Good for business, I suppose."

"I'd rather you didn't come here again," she said, very seriously. And from then on he respected this restriction. She never mentioned the matter again.

Standing in his kitchen, Harper was sorry he had even alluded to it. He had not meant to violate her secret or be mean, only to suggest the parallel. There were certain privacies one had to respect in this kind of relationship.

"You have your secrets and I have mine," he said as she sipped the coffee through the steam.

"I suppose they keep us together," she said, "make it possible to imagine certain things."

"I imagine," he said.

She smiled. She was not angry with him, nor he with her. Everything was fine, hidden and fine.

"You could take the day off," she said. "We could play."

"There are some things I have to finish," he said.

"There are some things we could start."

"I'm sorry. Duty before sin."

"Duty is sin," she said.

3

Donna Birch finished ironing the last of her husband's shirts and rummaged for something more to do. The plot of the police show on television had eluded her, and there was still a half hour before the next one would start. She pulled his summer uniform from its hanger in the closet. He wouldn't let her do his khakis or fatigues. Too much hassle, he said, all that starch. She pressed the creases of the green pants into blades and then eased the iron gently around the row of ribbons on the coat. Sometimes he seemed troubled that there were so few. But she was just as glad. He would have been decorated more if he had been sent to the combat zone, but she might have lost him there. She was thankful that the issue never arose.

She unplugged the iron and sat it on end to cool. Now she could turn the noisy little air conditioner on without blowing a fuse. She switched it to low and stood before it in the window, letting the breeze blow cool and damp through her blouse. The streets were empty, and still no sign of her husband returning.

Donna worried about him sometimes. Since being reassigned back from Japan, he had become so restless. At first she hadn't noticed anything special, what with all the unpacking and meeting new friends and all. Except, of course, that they did not have the extra money coming in anymore. But soon he had gotten another stripe, and they were OK. They were a whole lot better off than they had ever expected to be. The Army had been good to them. She couldn't complain.

That's what she always told him. She couldn't complain. The money didn't matter. But as the months wore into years in the new assignment, he became more and more dissatisfied, more and more withdrawn. He had started to drink at night, not heavily, just a glass or two of Bourbon and ginger. It was just that he had never felt he had to do this before. And he complained about the job. It wasn't challenging. Just a lot of paper. He might as well be a Spec. 4 company clerk, he said. The officers didn't have their stuff squared away. This one was incompetent. That one a little Napoleon.

Maybe he should try for a commission, she suggested. Some of his officers had encouraged him along the way to go to OCS.

"That isn't it, honey," he replied. "It isn't my rank."

Trouble was, she didn't really have any idea what was disturbing him. One of her girlfriends, the funny one next door, had given her a book about the phases people go through and said it would explain everything. Seven-year itch, her girlfriend had said, you'd better be careful. But they had already been married twelve years. The book was full of stories about couples breaking up. If it wasn't one thing, it was another. Sam J. was worried about getting older. Susan T. didn't feel fulfilled as a woman. Donna didn't know what to make of all this, even after she had read the book through to the end. Every age seemed to have its own excuse for sleeping around. But that was never a question between Jerry and her. It just never came up.

"They're all a bunch of he-men at heart," her girlfriend had said. "They keep seeing the young ones come in as privates. How would you like to have a bunch of eighteen-year-old sex kittens around you all the time? Pretty soon you'd be wearing hot pants and going braless just to keep your self-respect. Our men get tough because they think they might be able to scare age away."

But Jerry had never been that way. He had never tried to

scare anybody. He had always been quiet and gentle, even as a boy. He had shared everything with her, even his tears. There was never any question that she would marry him, nor that he would ask her. She would have done anything he wanted, but he had never pushed her to do the thing that wasn't right before their wedding night. Old-fashioned or not, that's just the way he was. And she was proud of it.

Then after quite a few years of marriage they had been assigned to Japan and had met the nice man from the European firm who had given Jerry extra work on the side. She had never understood what it was; too technical for her. It had meant good money, though, and Jerry had spent it on nice things—clothes, furniture, decorations for their place. It had also meant that he was away from her more, busy with meetings in the evenings. But she had tolerated it. He had seemed happy with whatever it had been that he was doing.

That was what she missed most when they got back to the States, having him feel good about himself. He sometimes became sharp with her now, not often, but it was so out of character that his wounding tongue hurt her all the more. He could be very sarcastic, and she didn't know where he had learned how. He would complain that the apartment the way she decorated it looked as cheap as a Japanese brothel. And when she would suggest replacing a piece or two or getting a new print to hang, he would remind her that they didn't have bags of cash to throw around anymore.

"They were always happy with your work for them in Tokyo," she had said one night. "Maybe we can quit the Army and go back there. Did you ever ask if there could be a full-time position?"

"It wouldn't be the same," he had said.

"You don't know that for sure," she said, touching his hand with hers and feeling his fingers stiffen. "It might even be better."

"I know what I know," he said. "I know what is and what isn't anymore. Don't press me, Donna."

"Sometimes it helps to talk things out."

"And sometimes," he said, "that's impossible."

So when, several nights before, the telephone call woke them up—a strange voice asking for him and then just a short conversation, no more than a few words—she was pleased to see him brighten up. She did not press him that night. But the next day she asked him about it. He was evasive, but she persisted.

"It's a good thing, Jerry?" she asked.

"Trust me," he said. And she wondered why he had put it that way. She had always trusted him.

Donna tried to pick up the thread of the story on television, but it was hopeless. The mesmerizing flicker of the screen did not capture her. She got up, put away the iron and board, rearranged some figurines on the shelf. Even though they had a spare room now just for relaxing and doing the chores, she couldn't stand to let it get disorderly. Finally she went to the sewing box and took out a piece of embroidery she had nearly finished. The pattern was intricate, old-fashioned. She liked the curved, intersecting stitches. They reminded her of her childhood on the farm where her mother taught her the craft of embellishment so that even plain things might be made to seem special. Jerry didn't like the fancy tablecloths and runners in their own house. They just didn't fit their life anymore, he said. So she sewed them as presents for the folks back home and as a way of using up time.

She held the piece up to the light by the corners. A needle dangled from a white thread. She examined the tiny circles and swirls. It just wasn't the same when you were doing it only to occupy yourself. Why didn't he come home? She let the piece drop into her lap. Her hands just weren't steady enough tonight. She wasn't good for anything.

Then a key rattled impatiently in the door. Glancing around the room to make sure everything was neatly in place, she put away the embroidery and went to meet him.

"You waited up," he said, obviously pleased.

"I wanted to hear all about it."

He hugged her, but not long enough. Then he headed for the kitchen. No, she did not suspect another woman, she later recalled. That was ridiculous.

"I think I have that job back again," he said, making himself a drink. "Would you like something—to celebrate?"

"That's wonderful, Jerry," she said. "Is it with the same people?"

"A little Bourbon maybe?"

"Is there any wine?"

"Sure."

"Is it the European firm again? I didn't know they did business here."

"Is this enough, or do you want a bigger glass?"

"That's fine."

He turned and handed her the wine, making a toast.

"It's going to be better now," he said. "I promise."

She smiled and took a sip with him.

"If it makes you happy," she said.

"It makes me feel that I'm not just wasting my time," he said.

"I love you," she said, and it troubled her to think that for him just the love was never enough, even though for her this was everything.

"I love you too," he said.

She did not worry about another woman, she insisted during the interrogations, and she had her reasons. She did not want to go into just why. But when we finally pushed her and she finally felt the need to prove that her faith in him, though she had not cheapened it by inquiry, was at least not misplaced, she told us what had happened later that night. They had sat together for a little while, finishing their drinks and watching television, letting it be their only point of contact. Then he had stood up and taken her by the hands. He had led her gently but urgently to their bed. And they had made love. Explosive love. Whatever other secrets he may have kept from her, she was as sure of this as she was sure of her husband's honor: he was seeing no one else.

Some of us found her loyalty not only credible but also commendable and—if we let ourselves lapse for a moment—really quite sad.

4

By now you may wonder who I am and how I presume to know these things: the question of sources and methods. Fair enough. But like all serious questions, this one must be approached obliquely.

I have made my life in the Agency. I have no name outside of it and perforce may take none here. On the organization charts the newspapers like to try to piece together and publish, I am one of the empty spaces, the question marks. I take my anonymity seriously.

Likewise, by the way, the anonymity of those whose story I am telling. I have given them names—Harper, Birch, Rubashov, Kerzhentseff—names with a certain resonance. But I am duty bound to protect some secrets, and their real identities are among them. I take this as a fortuitous metaphor. You see, these people are both false and true. They will be known to you the way they know themselves.

My work has always been in counterintelligence. And it was in the course of my employment that I became acquainted with this case, which I have given, for my own purposes, the code name CONVERGENCE. I have done an exhaustive study of the documents of the case, both as it evolved and afterward. Of the principal players, I had personally known only Harper previously. Despite his bias against my work, our relationship has always been cordial. For some reason he has not identified me with the warp of mind he believes afflicts counterintelligence personnel. Perhaps this is only in deference to rank and seniority. As for the others, though I have met few of them face to face, I have gotten to know them all very well. I have read every transcript, reviewed every tape. I have gone over the product of our surveillances (of this, I am sorry that I can reveal little more), and I have pored over all of Harper's own memoranda and notes both in the Agency and in his personal files. The Soviets, of course, are a different case. I would not pretend to speak for them. We could not question them. We have to speculate. But we are not without some data. Our dossiers on them are extensive. And they proved to be quite instructive.

But is my knowledge complete? Is what I present here properly thought of as history with the names changed? Unfortunately, intelligence work is antihistorical. By its stratagems it seeks to frustrate the truth as well as find it. Facts are contrived toward some distant, unwritten goal. An incriminating message may be forged. But was it created to reveal a truth that otherwise would have remained obscure? Or did we contrive not only the evidence but also the proposition it was put forward to prove? We are antihistorical because we act into history through falsehood. And it is our highest purpose to leave our complicity hidden and ambiguous.

More to the point as far as the question of sources and methods is concerned, we cannot adhere to the historian's discipline. "A perfect historian," wrote Macaulay, "must possess an imagination sufficiently powerful to make his narrative affecting and picturesque. Yet he must control it so absolutely as to content himself with the materials which he finds, and to refrain from supplying deficiencies by additions of his own." In intelligence work the "deficiencies" are precisely what we are expected to supply. What are the Soviets' unadmitted intentions? What weapons do they have that we have not yet spotted? Action cannot wait for certainty. Motive, deception, the things not seen or seeable, these are at the center of our endeavors. We do have techniques, a methodology. We create situations—experiments, if you will—which seek to draw fact out of the darkness. We apply the art of informed conjecture. But still we always deal in the middle range of probability. Our writ confines us to the realm of surmise.

At one time you may have been shocked by this lack of precision, this tolerance of error, these imaginative leaps. At one time you might even have read Macaulay. But now the writers you take seriously ridicule as naïve the aspiration for simple truth; they insist upon their right to violate all texts. The philosophers' unanswerable doubts have freed us to make up history whole, or in the name of critical inquiry to deface the poem of another man's mind. So you have become altogether used to the blending of fact and fiction. You have accepted the barbarous new words *faction* and *factoids*. You are comfortable with things that may not be the case but that will be treated as the truth simply because they ought, in some aesthetic sense, to be so. It does not shock you to read the interior monologues of the most taciturn individuals: the secret fears of Popes and Presidents, a condemned man's private epic, what went through the mind of a chief justice as he listened to an argument. You may not have the self-restraint Macaulay said was essential for historians, but you are fully prepared for espionage.

So when I fill in the gaps, recreate the conversations we did not record, embellish the characters' stories with all the finer details of life that are lost to the closest surveillance—the passing pain in Harper's hand as he let go of the knife, the sound of Birch's key in the door as he returned from the Hay-Adams —I trust you will not be put off. I vow that as to the essentials I have taken my characters at their words. (You, of course, need not feel the same constraint.)

As to my own motive for breaking silence, that is up to you to judge. I could list reasons: what the poet calls an old man's anecdotage; the desire to justify myself; even, perhaps, the need for expiation. But you do not have to know me. I can only reassure you that by writing I violate no duty, compromise no principle. It is, after all, not my story—except in the sense that it is all of ours, we who recognize that security must exact its price and claim its victims.

TWO
Tokyo, 1971

5

In the beginning, the Tokyo assignment was not meant to be a challenge for Harper. He was coming off a bad episode. His nerves were shot. And the last thing on anyone's mind was to hand him another dicey one. With some agents, the best antidote to a crisis in confidence is the fire. But Harper was not that kind of man. He had an active mind that needed time to sort out what had happened to him. In Tokyo the understanding was that he was to be bored glazy and carefully observed.

He was given responsibility for one or two moribund operations which had long since ceased to be productive and which required little more of him than a proper accounting of pension payments to a few played-out old men and a fair amount of paperwork that he and everyone else knew went unread. Harper hardly resented this, though he realized that his career was stagnating. He was not at all sure that he wanted to be outside at all. But it would have been unfair to bring him inside just then. The Agency was giving him every break. If he was to realize his potential—inside or outside—he had to regain his legs and prove himself anew.

Harper lived alone in Rappongi before he was married, and

every day he walked to the Embassy through the narrow, complicated streets. The city had been thoroughly rebuilt after the war. But it had grown back on the same labyrinthine pattern that had developed over the centuries. In much of the city the new buildings were anything but alien. Apartments, showrooms, office buildings without nationality rose on the twisting, immemorial streets. But what Harper sought out were the back alleys, forgotten in the rush of peace. He was taken by the economy of the small homes and shops, the way all space served many purposes, like the perfect word. A single room was bedroom, playroom, parlor. In warm weather the distinction between inside and outside blurred as people threw open their shops and set up stands in the sun. The Embassy itself was designed to fit in with such surroundings. It was large and walled, of course, but its lines were meant to give the illusion of smallness and simplicity. The Japanese buildings closest to it loomed up in stark, bulky concrete, unashamedly Western in contrast.

In his free time, Harper kept himself occupied with a course of reading, history mainly, things he had missed in school. He appreciated having the leisure to pursue the classics, and it was his way of trying to come to terms with his own immediate past. Though Janet remained in Saigon doing her own methodical work, they managed to get together from time to time on neutral ground and never mention war. These were the two sides of Harper's life in untroubled balance: the past made tame in works of literature, the present and perhaps even the future embodied in a pretty woman who had known him intimately enough in his troubles that she did not feel the need to talk about it.

It was because of the steadiness of this equilibrium, and not designedly to destroy it, that the station chief finally decided it was time to push Harper back up to a slightly higher pitch of risk and reward. Had he known how the Birch case would develop, the station chief could not have chosen better for his purposes, because at the beginning it looked to be little more than routine. In time the case developed as the Agency hoped Harper would develop—slowly but importantly.

Birch had not come directly to the Agency with his information. He had followed the chain of command. When the other side made its first illicit demand, Birch dutifully reported it to military intelligence officers at Camp Zama, who recognized the possibilities and informed the Agency, via Washington. Staffs came together at Langley and in the Embassy, layers mirroring other layers, and an agreement was finally worked out. It did not come easily. Those were still the days when lines of authority were purposely left unclear to encourage competition, the Social Darwinist theory of management. After several unpleasant sessions in which the military declined so much as to identify the potential asset, the real nature of the dispute began to emerge. We were concerned that Defense Intelligence wanted operational involvement to get a window on our most sensitive methods of collection; Defense feared that we intended to snatch the asset clean and cut the military out of the take except for whatever happened to show up under veiled attribution in our morning intelligence summaries. Once defined, the problem became surmountable. The station chief assigned Harper to work with the lawyers on language that would retain control in the Agency but guarantee Defense a piece of the glory should the asset prove productive. In the end all parties concluded the deal as amicably as creditors at a bankruptcy.

The asset in question was a staff sergeant, twenty-four, assigned to electronics work in a highly classified command center, married, no children. He had been through all the investigations and had a clearance sheet as complete as his shot record. He had graduated from high school before enlisting and had done very well in army schools. On his induction forms, Birch had listed his 4-H awards in his own careful, Kittle hand. He was a farm boy, the very safest kind. But Harper was unsure whether Birch would have the intellectual and emotional qualities he would need to play against the Soviets, for it was almost certainly the Soviets who had approached him. Harper explained to the station chief that he was doubtful whether a man of such a simple, exemplary background could really be expected to become comfortable with lies.

The station chief said he appreciated Harper's concerns, but he added that he was confident that human nature provided stores of deceit even in the plainest man. All we had to do was to find and touch them. And he added, for the record, that the right to terminate the operation was the Agency's, not Harper's alone. It was a decision the station chief reserved for himself. "This government is a topological wonder," he said. "Everyone, at every level, top to bottom, is in the middle at precisely the same time. If you want to bitch the deal, come to me."

Harper agreed, although he was wary about ceding control to anyone. He did not want to repeat his own mortal errors. And it was understandable, damned transparent even, to those of us who read his first uninformative cables on the case that he was holding back.

Harper chose to make his first contact with Birch at Camp Zama. He wanted to establish his own identity firmly in Birch's mind. If the case progressed at all, he knew he would be calling upon the man to do certain things that might give him reason to wonder. Harper felt he was going to have to bring Birch along slowly. This was no soldier of fortune. He was simply a soldier, and very often that was exactly the opposite. When the crisis came, Harper wanted to be sure that Birch did not face the terrible doubt about which side was which, what interest he served, whether there was any difference. We often find that our agents tend to see in their assets weaknesses that they have first discovered in themselves.

We paid close attention to Harper's reports on the formative stages of the operation. At Langley the consensus was that the signs were encouraging; Harper showed no inclination to balk. He had sound, tactical reasons for his actions. There were other reasons as well, which we discovered only much later in the course of our investigation. Sometimes the personal and professional imperatives tug a man in opposite directions, but in other circumstances, they all pull on the same vector. The psychologists call this "overdetermination," by which I guess they mean too much of a good thing. But I would call it cleverness, the ability to line up all the targets. Take, for example, Harper's decision to travel to Camp Zama on the GI shuttle. This had its practical logic. Embassy personnel sometimes rode the bus to avoid taxi fare. Its circuitous route through Tokyo would require a difficult rolling surveillance if anyone was trying to find out Harper's destination. Once he reached his destination there would be no ostentatious embassy car, no fanfare at the gate. He could come and go quietly. As for "overdetermination," it must be noted that the shuttle was the vehicle that took soldiers who had been on R and R back to the airport for the return flight to the combat zone. It must have seemed to Harper that, in a sense, he was also going back to war.

"They can waste me now, 'cause I've been to heaven and know it's fine."

"Well, we goin' back to the hot place first."

Harper sat in the bus silently, listening to the soldiers, remembering.

"If the Dinks are all the same, how come nobody tried to grease me here?"

"That was another war, fool."

Harper caught glimpses of their faces. Young, so young. That was the movies' biggest lie, to make soldiers seem ready.

"Back where I come from they say a man gets laid three times in a day is gonna live forever."

"Hell, if I'd of known it was that easy, I'd of paced myself."

Before the bus started to roll, its unmuffled engine relieving them of the need to seem strong, Harper looked out past the fence. Traffic moved in darting miniature. No surveillance team to worry about on the street. No one was out on foot either. The monument shop on the corner had no customers. The few cars parked within sight were empty. No eyes followed him, none that he could see. Ever since he had left Saigon, Harper had paid no attention to these signs. And now, without even thinking about it, he was back into the routine of the trade. The habit returned to him whole, as if the soothing interlude had never happened; his remission was at an end.

Camp Zama was just like the other military installations he had known. Drab, functional buildings squatted in open, orderly rows. Patches of dust were raked in long, straight strokes. Even the gravel was painted, piece by piece. Here and there some enterprising officer or NCO had tried to brighten up his area with a gaudy, ill-drawn sign or tiny bed of flowers. The effect was unspeakably sad.

Colonel Robertson, Birch's CO at the communications center, turned out to be a genial and accommodating sort, though he seemed a little intimidated by the proceedings. Accustomed to dealing with the conventional kind of secret that required no more wit than the ability to keep silent, he did not know exactly what to make of this new arrangement. He repeated Harper's false name—a little too loudly, as if he felt he should spread the lie as widely as possible—then shook Harper's hand. He led Harper to a plain little conference room where Birch was waiting. Then he left the two of them alone.

Birch was by no means an imposing figure as he stood there, hat held in front of him like a codpiece, waiting to be told to sit down. His face was one smooth curve, hairline to chin, with nothing to distinguish cheek from forehead but location. This, along with the dense brown hair trimmed high above his ears, gave him a boyish appearance. It was not precisely a baby's face, because even infants have expressive mounds and hollows with which they show their joys, their hungers. Birch's face was smooth, like an egg, and Harper had the feeling it would reveal to the observer anything the observer thought it should.

"I'm Jerry Birch, sir," he said, sticking out his hand to be shaken.

"Richard Harper. The other name was to protect us both. Why don't you have a seat."

"It's my real name," said Birch.

"I would have assumed as much," said Harper. "We ought to be telling one another the truth right from the outset, don't you think?"

"I mean, some people think it's Gerald or Gerard," he said. "But my parents actually named me Jerry. That's the way they put it on the birth certificate. They didn't want me to be saddled with a nickname."

"What do your friends call you?" said Harper, deadpan.

"Why, Jerry," he said. "They call me Jerry. Like I said, it's my real name."

Harper smiled but did not push it. A literal man was all right. In espionage you did not have to have a taste for irony to see everything double. And doubled was as simple as it came.

Harper began by revealing to Birch that he represented the Agency. He explained that military intelligence had agreed that the civilian organization should take the lead in this case and that Birch's superiors had been informed. Birch was to report to Harper and no one else.

"I know you appreciate the seriousness of this business," Harper said.

"I guess I do."

"It will, I think, become clearer as we go along. I want you to begin by telling me about your first contact. Our time here is limited, but I think it is important to review the background."

"Whatever you say, sir. I just hope I didn't do anything wrong."

"Why should you think you did?"

"Well, they don't give you rules for this kind of thing," said Birch.

"There aren't any."

Harper watched Birch's face for a sign that he felt the true weight of those words. He thought perhaps he saw an instant of recognition, but he could not be sure. It was the perfect face.

"I first met the man casually," Birch said, consulting his lap. "My wife and I were out on the town. We stopped in a bar after dinner, and he came up beside us and started a conversation. He looked familiar. You know, I thought maybe I'd seen him around, but I couldn't place him. It was like, what's the word for it, a French word, when you imagine you've seen something before but can't exactly be sure. What do they call that?"

"Being under surveillance," Harper said.

"No," said Birch. "It's another word. Anyway, he had a funny accent. He wasn't Japanese. A white man, you know? We were the only ones in the place, so I didn't think anything of it when he was friendly. He bought us a round of Suntory, the kind that's supposed to taste like Bourbon. Then I bought him a round. He said he was a businessman representing a European trading company. I told him what I did. The Army, I mean. Nothing specific.

"I kind of expected that would be the end of it. I mean, most people aren't too interested in soldiers. But this guy was different. He kept me talking, even brought my wife into it. She liked that. One of the good things about overseas duty is that you meet so many different kinds of people."

It was too good to be true. Literally too good. There might be innocents abroad, but they wouldn't be in the Army. Even the most naïve soldier has eyes. Why did he think they taught him how to fire a rifle?

Yet at the same time, Harper found himself tempted to believe in this man. Perhaps it was Birch's utter lack of self-consciousness. Perhaps it was that smoothness of countenance, the absence of care. If Birch could be trained to control this gift, put it in service of certain purposes, he could be formidable.

Birch went on to tell how he had given the man his name and home telephone number and taken the man's in return. The name the man had given was Nowicki.

"Did you give him your number at work too?"

"No, sir."

"Why not, Birch?"

"I didn't think it would be right."

"So you did have doubts about him."

"No, sir. Not really. I just figured that my work was my business."

"Maybe you didn't want him calling you where others might overhear."

"I told the colonel about him the very next day," said Birch. "I didn't care who knew. I told the colonel we'd gone to Tokyo and met a real interesting guy. You can ask him if you want."

"Where did you think this Nowicki came from?"

"He said he was a citizen of Europe," said Birch. "So many wars. So many displaced persons. Nationality didn't mean very much anymore, he said. You were born in a place, sure, but the map kept changing."

"Have you ever heard a Russian accent?"

"In the movies. Greta Garbo on the late show. Or is that German?"

"Did you notice anything special about his accent?"

"He spoke real good English. I didn't have any trouble understanding him or anything."

Harper stood abruptly. Birch followed him up with his eyes. Curiosity, Harper thought, not fear. But it was hard to tell.

"Go on then, Birch. You might as well go on." Harper wanted to hone him to an edge, but Birch remained easy, trusting, secure.

A couple of weeks had gone by, he said, before he got a telephone call from Nowicki. He had the precise dates, and they squared with the earlier reports. Everything did, down to the Japanese whiskey that's supposed to taste like Bourbon.

"Are you sure of those dates, Birch?"

"Yes, sir."

"Did you write them down? Was there some reason you made a note of them at the time?"

"I'm very good at remembering things. Names. Numbers. They just seem to stick with me, at least when I want them to. My wife says it's a gift."

"All right, Birch."

Nowicki wanted to take Birch and his wife to dinner. Birch agreed on condition that Nowicki promise to come to their house for a home-cooked meal sometime. It was settled. They got together twice, the three of them. The talk was small and unremarkable—Birch would have recited it word for word if Harper hadn't hurried him along—until Nowicki noticed on Birch's shelves a row of electronics books. He asked about them.

"I told him it was my job," said Birch. "We talked about it

awhile. You can be sure I didn't get into anything classified. Don't get me wrong either. It wasn't as if he pushed. If he had, maybe I would have caught on earlier."

The third time they got together, Birch came alone. His wife had caught the flu and insisted that he keep the appointment. Nowicki took him to a French restaurant on top of the Okura Hotel. They had a view of Tokyo's tower and relaxed with drinks to the sound of a flute and harp duet performing on the other side of the room.

"When I saw the prices on the menu I told Mr. Nowicki it was too expensive," Birch said. "I could never repay him, not on an E-6 salary. He just waved his hand and said it was all on his company. Anyway, he said, you couldn't get home cooking like my wife had served at any price. But it must be hard, he said, getting by—with Japan so high. I told him about the PX. I guess I even offered to pick up things for him there if he wanted, American liquor, that sort of thing. He went on about how tough the money situation must be for a man like me who appreciated the good things in life. I told him I didn't really have champagne tastes at all, thank God, but if they promoted me to E-7 I wouldn't turn the extra money down. 'Taste only shows itself when there is an opportunity,' he said, and asked if I'd thought about picking up some extra pay on the side. I told him there wasn't much chance for that overseas the way there was in the States. At Fort Meade I used to do a little TV repair, but not anymore."

"Did he show any interest in your assignment at Meade?"

"I'm not sure I even mentioned the place. We talked TV."

"I thought you had the perfect memory, Birch," said Harper, and he turned his back on the man.

"I did mention Meade that night, but not just then," Birch said, raising his voice. "I told him about it."

"About security agency school?"

"No, sir. That never came up."

"But you had the impression that he knew."

"Not at the time, no. I don't think you understand the way he acted."

"Oh, I think I might," said Harper, turning back to face him. "It isn't really so complicated, is it?"

"I should have seen something coming, you mean," Birch said. "It just didn't cross my mind."

Harper could have relieved his anxiety, but he didn't. He just shrugged his shoulders and told Birch to go on.

"A few days later," Birch said, "he called up to say his firm needed someone to do a little electronics consulting work. Consulting work is what he called it. I told him I didn't know anyone in that business, but he said he meant me. I told him that I was no expert, but he said he was confident I would meet their need. It was fairly elementary stuff, he said, not worth bringing a technician all the way from European headquarters. The money would be good, he said. I told him I didn't mind money so long as I earned it."

"Admirable, Birch."

"I can't help the way I am."

"But what way are you?" Harper demanded. His voice was loud, and Birch flinched. "That's what's puzzling me," Harper said in a softer, more menacing tone. "What way are you?"

"I don't know what you mean," said Birch, and it was an apology.

"I guess we'll find out together. Now, please go on."

Nowicki set the hook slowly, gently over a period of a month, paying Birch several hundred dollars for answers to a series of innocuous questions. None were military-related. And Birch had not even felt the barb. He let himself be persuaded to buy some expensive stereo gear on time to "live up to his income," as Nowicki put it. Eventually, though, Nowicki started to take the slack out of the line.

"He wanted me to get him some specifications that were classified. Nothing really sensitive. Confidential is all. I don't know whether I'm supposed to talk about it with you."

"I've read the reports," said Harper. "I know what he wanted. Don't worry. I'm cleared for it."

"It's just that you're a civilian and all."

"And I don't have an accent or a wallet full of cash."

"That's not fair," said Birch. "It surprised me when Mr. Nowicki asked me. I was nervous about him too."

"Did you tell him the material was classified?"

"Yes I did."

"Good," Harper said. He sat down at the table and leaned forward on his elbows. "Did you refuse to give it to him?"

"I tried to," said Birch. "But he wouldn't hear of it. He said the firm was bidding on an American contract and just wanted to have a fighting chance against the established competitors. There was nothing wrong in the firm's having the information, he said, since it was all going to stay in the family in the end anyway."

"What did you tell him?"

"I thanked him for the chance to earn some extra money, but I told him that I didn't know the answer to his question. 'But you could find out,' he said. I told him it might take awhile. And anyway I wasn't supposed to discuss these things with anyone who isn't properly cleared. He said there was no hurry. I should take my time and decide. He gave me some more money."

"How much?"

"I tried not to take it, but he insisted. He called it good faith money, a retainer."

"How cheaply do you sell yourself, Birch?"

"I didn't sell anything. He gave me a hundred dollars. I still have it. He gave it to me in green. Look, I'll turn it over to you if that will do any good."

"Altogether, how much are you into him for?"

"They weren't loans. I don't owe him."

"Oh, you owe him all right. As far as he's concerned, you're mortgaged up to here. And now he wants the rest."

"He paid me $325, counting the last hundred," Birch said.

"What if I told you I believed it was more?" said Harper.

"I can add," Birch said. "I keep track . . . Unless you mean the meals. It's more if you count the meals."

"Why hang yourself on such a small point? It might have been a few hundred more that he gave you. You can't afford to turn it all back to us. We understand that. We're interested in the truth, not the money."

"I'm telling you, Mr. Harper, he paid me $325 exactly, plus a few good dinners."

Hard as Harper pushed him on this point, the man would not budge. It was a good sign, his firmness. But you could not always be sure whether a man like Birch could learn to be as solid in a lie.

"Why did you wait so long before reporting your contact with Nowicki?" Harper asked. "You took money, maybe three hundred bucks, maybe more, for feeding information to someone you had no reason to believe was who he said he was. Then suddenly you got scared."

"He had never wanted anything classified before. I wish I'd been smarter. I wish I had never met Mr. Nowicki, never took a penny. Nothing like this has ever happened to me before."

"He had a Slavic accent, and he was paying you large sums of money after learning that you work in the security field. Please, Sergeant Birch, I've seen your IQ scores. I've seen your record. You are not a stupid man."

Harper had been building to this one last test, and now he watched his subject carefully to detect any small signs of weakness that could cause them problems as they went ahead. He was not looking for evidence of falsehood, for he believed Birch was telling the truth. He was looking for the qualities of toughness and plausibility that his asset would need. Birch met his eyes, not pleadingly, but sure.

"I'm sorry I didn't report this earlier," he said. "I truly am. But at first there was nothing to report."

"You are asking me to believe that you are terribly gullible."

"If that's what you would call it, then I guess I am."

Good again. He was unbending on the facts but not defensive. He wasn't arguing a case. He was simply stating what he knew. Harper was excited by his potential.

"I guess this ruins my career," said Birch.

"Changes it," said Harper.

"I'll lose my clearances."

"You'll be needing higher ones. There will be an investigation. I hope you aren't hiding anything."

"I thought this was the investigation. I don't understand."

"Just routine. You've been through them before. It won't involve your contact with Nowicki. We're going to improve upon that. That's our property now. I want you to call Nowicki, or whoever he is, and tell him you'll try to get what he wants."

"It's illegal."

"Sergeant Birch, we intend to use you against this man and whomever he serves. It will take some time. You will have to be trained. It will not be without risk. No one may know of it but you and me. Not your superiors. Not your wife. You will be living a lie."

Harper gave him no opportunity to balk. You do not ask a man's permission to warp his life beyond recognition. You state the requirements. They are, at the time of recruitment, abstract and even somewhat exciting. Their true burden only becomes apparent when it is too late to reconsider.

"Do you think you can learn how to lie?" Harper asked.

"Lie to Mr. Nowicki?"

"Lie to everyone except me."

"It will be good for the country?"

"The higher good," said Harper, and he disliked the cynical way it came out. "Always for the higher good."

"I should tell you that I'm on levy to Vietnam. I expect my orders to be coming down any day."

"That will be changed."

"You can do that?"

"We will have it done. You've never been in combat?"

"No, sir. This would have been my first chance."

"You won't resent missing the opportunity, I trust," said Harper.

"I'll do whatever you think is best."

"You won't be getting out of danger, you know," said Harper. "You will be in a kind of combat here. It will be more personal, more intense. But there is no glory in it."

"I just want to be useful," Birch said. And Harper remembered those same, unremarkable words chiseled into stone in

the New Haven courtyard, attributed to Nathan Hale. Maybe
they were as eloquent as any could be. Maybe that was why
someone memorialized the words in New Haven and built a
statue to the man at Langley. You commit yourself to the util-
ity of your country. You agree to become a piece that is moved
about by others, to be made into whatever persona the moment
requires, to surrender all claim to identity. Maybe eloquence
would be a lie. *I wish to be useful.* It could be the last, small,
unadulterated truth a man ever utters. And if so, the words
were right for stone, an epitaph.

6

In the beginning, the relations between asset and handler are always somewhat awkward. The case agent is a warlock initiating a neophyte into a secret sect. Special signs are exchanged to establish identity, incantations to warn of danger. The rites of the dead drop are explained: You shall go to the appointed place and leave your message there unguarded in perfect faith that an unseen acolyte will be there to receive it. This is the fellowship of Orpheus without the love or music. There is no looking back.

Harper led the novice slowly and painstakingly. He made sure that Birch understood that no action was to be taken without his orders. He gave Birch a special code word that only the two of them were to know. Not even the station chief was privy to it because Harper insisted on absolute control. For his part, Birch did not seem to know what was serious and what was jest. This is natural, since everything was so strange and new to him. He likened the code-word discipline to children's play. Simon says pass the secret. Simon says use the dead drop. Touch your nose. Harper assured him that it was no game.

At the same time, Birch had a need to establish himself with Harper. He treated him as a confessor and went on and on about his past. Harper listened because he knew the loneliness. Birch was entering a universe in which he could share his experiences with no one else. In such an empty place, a man wanted to establish one saving bond, one connection that would abide when all others were violated. In such circumstances, a man becomes particularly vulnerable to the corrupting lure of trust. And though perhaps he, too, was prey to it, Harper tried to steel himself so that the connections all went one way. He wanted Birch bound tightly, but he needed at all times to be able painlessly to cast him off.

This was not because he found any reason to doubt Birch. It was only a matter of principle, for Birch seemed to have none of the usual weaknesses. As for money, he had no great concern for it. He had turned over to Harper almost all of what the man who called himself Nowicki had passed him. Harper had to insist that he take a regular stipend from the Agency in order that he keep up the appearance of living beyond his means. Birch's personal life was droningly loyal. From the pictures Birch showed of his wife, she was attractive enough, though not quite so magnificent as he described her. Birch was not the sort of man to ask himself what he is missing. In fact, he seemed incapable of disillusionment. Even as he was being tutored in deceit, he showed no inclination to question the methods which violated every American standard of fair play. He simply accepted them as necessary because Harper said they were.

This was something I picked up on early as we reviewed Harper's cables back at Langley. I was encouraged because it has been my experience that a man can continue in this trade effectively no matter how distasteful he finds some of its imperatives so long as he recognizes the legitimacy of its purpose. Let philosophers trouble themselves about the interpenetration of means and ends, emotionally the connection is unseverable. A man becomes dangerous when and only when he can no longer believe in the unique virtue of the larger mission. Whenever I detect the slightest evidence of this in an individual, I mark that man down as a risk forever after. This is harsh. I admit that. But it is my work. What are flags and anthems against the witness of the void? People do not turn because of money or drugs or lust or any of the other useful vices. These are only the media of a more fundamental failure. Men become disloyal only when, for whatever reason, they cease to believe that the side they take makes any difference.

From what Birch told of his boyhood in Cleanthe, Illinois, there was nothing to indicate that it had prepared him to entertain the corrupting doubt. Though the little farm town was dreary as dirt, Birch had words to praise even the soil. Harper pictured Birch's father as a small, hard man in a gray straw cowboy hat and jeans, fastidious and yet prepared to wade through manure when the work required it. Birch himself was also slight, though he claimed he was bigger than the old man. All the children were. But Birch had not been tempered by the sun as long. His toughness was not the same. It had not yet been weathered.

Their house was a white clapboard set back about a mile from the road and, unlike so many of the other farmhouses, kept immaculately painted and landscaped. Birch's mother was of the firm opinion that folks should live a cut above the livestock. It was she who climbed the ladder every third year to scrape and splash the planks with fresh whitewash, she who played foreman to the brood, assigning chores, supervising the husbandry of lawn and flowerbeds. Her boys and girls alike started out on the miniature tractor mowing the grass when they were barely able to reach the pedals; it prepared them to graduate, once they grew tall enough, to the big John Deere in the endless fields.

Cleanthe was little more than a stretch of railroad track bordered by a couple of buildings you would barely be able to distinguish one from the other as you whipped by in a streamliner. A grain elevator, a two-room schoolhouse, a grocery store with a warped screen door advertising 7-Up, and a dance hall where they served whiskey by the glass during the years when it was legal and suffered it to be brought in in Mason jars during the years when it was not. The trains threw off the mail in leather bags and picked up the outgoing with a flying hook that snatched it from where the postmaster hung it on an iron pole. Cleanthe's contact with the rest of the world was just that fleeting, and no one in the town ever complained.

Birch went to school for eight years in the red brick school-house, four years in one room, four years in the other. Every August someone came to sand and revarnish the floors. Opening day the place shone with expectation and smelled fresh with spirits. By the time of spring planting recess, the floors were ground down to a dull gray by the scrape of chairs and the scuff of muddy feet, and everyone was ready to take the next lesson from the land.

In such close quarters it was hard for a child to have any secrets, and it was not much easier for the adults. Telephones came to Cleanthe early, but the town was not big enough to have its own exchange; that honor was reserved for Welton, down the tracks. Still, every family had its own unique ring, and the etiquette was that Weltonians never eavesdropped on Cleantheans, and vice versa. This was a matter of sovereignty and decorum because it was easy to learn the trick of picking up the receiver to listen in when someone had died or had fallen newly in love. You had to wait until the ringing had stopped and then lift the earpiece ever so gently so as not to make a warning click. Of course, everyone could tell when someone else was on the line overhearing; there was a special hollowness and the rustle of fingers on the mouthpiece. But this was not espionage. It was a form of kinship. No one had anything practical to gain from it, except perhaps the preacher, who found it provident to stay current with the sins of his flock. It gave his Sunday sermons a certain hell-fire immediacy.

When it was time for high school, Birch went off to Welton, four miles away. Sometimes he hitched the freight where it slowed down for the switches. Most of the time he simply walked. The high school was enormous to him. It graduated nearly fifty every year. Birch did right well there, he boasted. One of his teachers had even encouraged him to apply at Urbana, but everyone knew that she had come from out East somewhere and had a lot of funny ideas.

Birch played basketball, came to fancy his wife-to-be that way, for she was a cheerleader. She jumped so high. One year the team even made the Sweet Sixteen in the state tournament. The papers up in Chicago wrote stories about it, calling the school "Wee Welton" for some reason. The team was defeated on the first rung of the ladder by a school from the city composed of the biggest seventeen-year-olds Birch had ever seen. But when the team returned home on the buses, the folks turned out anyway—late at night—with banners and a band, just as if they'd brought back the prize.

Before he left Cleanthe for the service, Birch had never owned a car, never drunk more than a couple of beers, and never used cigarettes after the first furtive times he had sneaked smokes with some of the other boys behind the barn when he was thirteen and disliked the taste. He earned attendance awards at Sunday school, participated in the 4-H. Unlike a surprising number of others, he never got a girl in trouble, though with Donna he came close enough to it in their favorite meadow that he felt like a hypocrite for not taking the side of his unluckier and more daring friends when the older folks started clucking their teeth in disapproval of weddings of necessity. He was not troubled by any other moral dilemmas, more for lack of occasion than anything else. Cleanthe lived by a rather straightforward—though unwritten—code. Departures were rare and never the subject of debate. Even violators accepted the legitimacy of the rules they broke. It was not so much intolerance that kept Cleanthe this way, during Birch's early youth before television lit out across the plains like the devil's missionary, as it was sheer distance from the sophistry of the more advanced centers of progress and decay.

The Birches had more contact with the outside world than most. They went up to the city about once every other year, flagging the silver streamliner down just past the grain elevator and riding it north with the mail. In Chicago they always stayed at the Old Grainman's Hotel on LaSalle Street, not realizing how much it had gone to seed since Birch's grandfather first took his young family there for an adventure. To them the close quarters, peeling radiators, and smell of mildew in the closets were just the way city folks lived.

Each trip, the Birches visited the Board of Trade, where the profit of their labors was daily gambled. Young Birch was fascinated by the screaming activity in the bean and wheat pits below the observers' gallery, by the men in their faded smock coats the color of half-dried hay wildly communicating with one another with raised fingers and raised voices, making money in some mysterious correspondence to the way the Birches made grain. The family ate in restaurants where the waiters brought the meals to the table already dished out on individual plates, and the boys did not get larger portions. They walked among the buildings too tall to get a look at whole, and Birch's feet grew sore from the cement.

No, the city did not lure Birch away from Cleanthe. Military service was for him a duty, not a deliverance. He could have stayed in Cleanthe for a lifetime, making occasional trips north to reaffirm the advantages of the soil, and dreamed of little more.

But, as his father said, the land carries a tithe. For some generations, like Birch's father's, it was exacted nearly universally. For others it never came due at all. But this was not to be seen as unfairness. It was simply the way of circumstance, like the rain fouling a harvest or a blight on the corn in the summer. The elder Birch had gone to Patton's Third Army in Europe and experienced certain things that he made it clear no longer troubled him at night but that still were not to be discussed. Mrs. Birch had waited at home, managing the farm as best she could with a hired man and Birch's older brothers.

This did not make the Birches extraordinary patriots. It is important to understand this point, for if they had been, perhaps young Birch would have come to question his father's sense of obligation. But the elder Birch's view was the rule; to doubt it in Cleanthe would have been as otherworldly as to question a man's need of a nose. And when Birch returned home on leave after his first reenlistment, even though it was during the bad years of the war, it was like coming home from the state basketball tournament. Cleanthe welcomed its boys home, win or lose.

Harper envied the steady compass of Birch's childhood. Harper himself had grown up with a different sort of privilege in a family with a quite distinct conception of duty. Duty was a universal solvent. It excused everything. Personal profit was its own morality, an indicator of the common weal. Advancement was a measure of service. Not that the family recognized no claim of society. His father, after all, sat on all the goodly boards and lectured his son on the importance of charity. But this was dependent, not automatic. The tithe came only after the good harvest. There was never any question of sacrifice.

Home and family played less of a role in Harper's childhood than in Birch's because it was so early supplanted by the first of a series of boys' schools. Home was for holidays, and this was explained to Harper as an education in self-reliance. The boarding schools he attended, as later in life his college, all had the same attitude. Their responsibility was to produce an uninterrupted flow of young men of a certain stamp to regenerate, without radically altering, society from the top. While Birch was learning obligation, Harper was indoctrinated into the more stratified sense of *oblige*. And by the time he went off to the university, Harper had developed from his special relationship with his family a quality of mind marked by civility, taste, mild skepticism, and detachment.

New Haven was meant to reinforce these things, though it did so by first challenging them. He forever remembered those four years by the sweet smell of wood burning in the stone fireplaces and the way it overcame the salty, maritime wetness of the wind off the nearby sound. He studied literature because a man should not need to prepare himself for business. He became interested in the theater and jazz. But you could only sit in the Greenwich Village coffeehouses listening to the icy jazz and spontaneous humor for so long before realizing that these improvisations would only carry you a certain distance. You had to do something, something authentic. By his senior year, Harper was ready for our approach.

We recruited many like him during those dull and dangerous years. They had been exposed to all the alien ideas, and this was a kind of process of immunization. The antibodies were potent in their blood, and we had reason to believe the protection would last a lifetime. They called themselves alienated, meaning that nothing suited them. But most, like Harper, really only rejected what was most ordinary about themselves. They disdained and feared all the traditional paths of attainment with which the futures they imagined seemed rutted. Yet they performed well in school, if not by their brilliance, by their discipline. They were afraid of boredom most of all (though when they met it head on in the tedious details to which even we at the Agency find ourselves routinely attending, they discovered it was as comfortable as a soft old chair). They had all toyed with what was then known as Bohemianism, for anything native was suspect, even a name. They gloried in the strange new poets, started doomed magazines, defended the interminable French. They listened to the composers nobody played. Politics meant little to them, not because what prevailed failed to square with the radical tenets they harbored, but rather simply because it seemed so graceless.

We made our approach to Harper through a professor who was one of his closest friends on the faculty. He happened also to be one of ours, though this would hardly have been suspected. He was considered very odd and was looked upon with distrust. The local veterans' groups often complained about his courses on Soviet politics because it was his habit to laugh rather than to sneer. Russia's foibles seemed to him infinitely amusing. And, since he had tenure, so did the VFW's.

"Did you hear the portly peasant's latest?" he asked as he poured Harper a sherry from a decanter which, in this progressive college, could be displayed prominently on the mantel of his office fireplace.

"Another premature burial?" asked Harper.

"He claimed Russian credit for the invention of the automobile," said the professor. "It was in *Pravda*. Another bit of revisionist history. Some of our scholars must envy their . . . well, their latitude."

"Typical," said Harper. The wine was like a warm breath on his tongue. He settled comfortably into the big couch of leather and buttons, and he thought there was no setting more sophisticated in its disdain. He admired the professor for recognizing that the real problem with Communism, as practiced, was not that it was classless but that it was so déclassé. The fat little farmer, Khrushchev, was the personification of what was wrong with their system—a man who made even mumbling, stumbling Ike seem like Pericles.

"Yes it is typical, isn't it?" said the professor. "Of course it is nothing but *vranyo.*"

"I suppose," said Harper, who had learned enough in four years of higher education to be noncommittal when you did not know the word.

"When Texans boast, they want to amuse. When the Russians engage in *vranyo,* they want to be suffered. You must neither believe nor contradict them, and you must pretend to take them seriously. *Vranyo* is a trial, like warm onion vodka straight."

"The man is an absolute horror," said Harper, and the rich sherry and tickling fire made the thought a comfort.

"Actually, though you are too young to remember it, he is a relative horror."

"Stalin, you mean. Sometimes I think he was, in a sense, more honest." Harper was well aware that, in a sense, anything could be made to be anything and that this was the way an open mind preened.

"I suppose tastelessness can be a form of violence," said the professor, flattering him by elaboration. "But while *vranyo* is a bore, it is benign. On the verge of a famine, they'll boast about their agricultural methods. Did you know, by the way, that a *khrushch* is a beetle, a farm pest? They've expunged the word from their dictionary."

"Fools."

"Of course they are not fools," said the professor. "They are much more dangerous than that. It is the difference between *vranyo* and *lozh,* the lie. Richard, they are the world's most cunning liars. It is a cultural trait. They can even lie themselves into belief. That, after all, is the necessary precondition of their ideology. And the Russians are as committed and convoluted as the medieval churchmen. If I make light of them it is only so that we recognize what is appropriate to fear—not their cant but their cannons. We have to be strong and very, very bright. It is work you really ought to consider."

"The Army?" said Harper. "I don't think I'd exactly fit in."

"Certainly not," said the professor. "For a young man like you there are much more efficient uses. I am talking about intelligence work."

"Spying," said Harper.

"I hope you will consider what I have suggested as confidential."

Harper was intrigued and flattered by the prospect of sharing with his elder so important a secret, and he agreed. Ours is the only trade I know in which recruitment is a form of on-the-job training.

"There are some people I would like you to meet, Richard. They are looking for bright young men, and with your permission I will give them your name. I think you will find them unlike what you expect. They could be you or I; they do not stand out. But they are terribly clever, and I think you would find the continuing education among them quite a challenge. Intelligence is a form of scholarship, you know."

Harper eventually went through the formality of indecision —several meetings with our team, two further conversations with the professor late into the night—but from the very beginning, I think, the opportunity took hold of him. As he walked under the old stone arches of the quadrangles or sat at the heavy wooden desks in the library, this prospect of a future gave the academic process of inquiry a newfound shape which, because it was secret, was proof against the common contempt.

I have always thought it unfortunate that in our recruitment we place such emphasis on the challenge of penetrating our adversaries' duplicities. It creates certain expectations that are not generally met. The Agency is not the academy. It does not engage in the higher forms of speculative thought. The real scholars do not last long with us because they cannot reconcile themselves to the fact that unriddling the other side's deception is no more important than perpetrating our own. And we end

up with too many men like Harper, who consider this side of
the work an unfortunate compromise, and too few like Birch,
who take to it with an unreflective zeal, like a prodigy to his
fiddle.

7

From the very start, Harper had recognized Birch's gift, but it was not until they began preparing the book against the man who called himself Nowicki that he realized how special it was. The plan, at the beginning at least, was very simple. Birch would copy out the information Nowicki wanted in his own hand, complete with the incriminating classification word CONFIDENTIAL. Then he would set up a meeting and turn it over. As they worked out the details, Birch zeroed in on the matter of tone. Harper let him play, encouraged it. Though the design was to let the man think he had Birch firmly in jeopardy and ready for blackmail, Birch suggested that perhaps he should say the document was easy to get.

"See," he said, "I could take the line that, hell, they don't even try to protect this stuff, so it couldn't be much."

Harper toyed with the idea. It would make a clever agent think Birch was trying to rationalize his treason already. It would make Birch not only seem ready to cheat his government but also ready to lie to himself. The deftness of this feint impressed Harper, for it was grace in elaboration that made a fiction work. Embellishment distracted the eye. A small turn here, an arabesque there. If you could find an element which at first glance seemed incongruous but which upon closer scrutiny fit perfectly, this was the highest art. But Harper decided to keep it simple this time. He did not know how sharp the man who called himself Nowicki really was. You must not throw a fake that is quicker than your opponent's ability to react to it. It was possible to queer the scheme by your own cunning. In the beginning it was always safest to be prepared for the possibility that your mark is a fool.

Harper told Birch that he was to say the document was kept in a high-security safe.

"You have access to it during the day," he said, "but at that time you cannot risk going to the photocopier. Too exposed. So you copied the data by hand at your desk. He will probably press for true copies. You will resist, become defensive. 'What am I, a notary public?' That sort of thing. Tell him that at night the safes are locked and alarmed. You can open them, but the breach is recorded on a pen register. You must have a reason to go into them because there is always an inquiry. He may not show it, but he's going to think you have the keys to the treasure chest."

"By God he will," said Birch.

As they rehearsed the story over and over, what struck Harper most was Birch's delivery. He seemed born to the work. It was as if the story they had invented had become for Birch the greater truth. Even at the time, however, Harper's enthusiasm was tempered by a kind of sadness that the skill came so naturally to a man so decent. And it did cross his mind that if Birch could so completely eliminate the distinction between fabrications and the truth, it might at some point become difficult for Harper to sort them out himself.

It was not long after the last rehearsal at the safe house that Birch made his telephone call to Harper, as arranged. The man who called himself Nowicki had made contact, and details of the meeting plans would be left at the dead drop the next day. It was not, strictly speaking, necessary to go to such extremes of security so early in the operation, but Harper wanted to get his asset used to the discipline and to develop the sense of blind trust that might become essential later. Just as he had made Birch repeat their special code word often at the safe house with all recorders off so that it would become established as an incantation of recognition and obedience, so also he used the dead-drop technique to develop Birch's confidence (and fear) that even when Harper was nowhere to be seen, he was still watching.

The cemetery he chose for the first drop site was on a drab little hill. Surrounded by a gray stone wall with only one entrance, it had an obvious advantage, since Harper wanted to check to be sure Birch had not picked up a tail. Nearly an hour early, Harper positioned himself in a little eating establishment which had a clear view of all approaches to the gate. Seated alone in the shop, facing the bright, streaked window, he read at a copy of Christopher Marlowe a sentence at a time to avoid missing anything on the street. He found it hard to keep track of who the hell was the devil.

The pastry was quite good, and the middle-aged proprietress seemed dedicated to keeping his cup of tea nearly too hot to drink. Eventually a local couple wandered in and diverted her attentions for a time. The three of them jabbered away excitedly, and Harper wondered whether they were talking about him.

Impatience is a weakness in a case agent, so when the appointed time passed with no sign of Birch, Harper simply ordered another dish and read on. Miniature trucks strained up the hill. Old men and women took their morning constitutionals. Children in their bright, starched uniforms marched to school.

After forty-five minutes Harper made his move. Birch had been warned of the need for close timing. Harper was disappointed in him and upset that the mission was getting off to such a ragged start. It was standard procedure to check out the site even in these circumstances, so he hurried across the busy street and made for the gate.

The cemetery was shaded by the walls and by some sparse trees. Gravel paths wound among the graves, each marked by a tall wooden stake bearing a row of characters. They didn't chisel names in stone. Like an individual mind, an island had room for only so many memories. The names faded in a generation and then were replaced with testimonials to other lives. The only permanent monuments were the small shrines on their pedestals. Inside them, families had placed sticks of incense, prayers to the dead. Back in the States you would sometimes see these graceful pieces decorating somebody's back lawn or garden, where they often as not held scraps for the birds. Harper half-expected to see some Japanese patio graced by a heavy tombstone and cross. They admired inappropriately, just as we did.

Harper followed the path to the farthest corner. Behind a bush, hidden from the eyes of any idle passersby, stood the shrine Harper had chosen. He walked across the grass to it, not knowing the protocol for avoiding the graves, and quickly pulled the leafy branches away from its opening. There, to Harper's surprise, a small piece of paper lay folded neatly on the altar. He could see that the words were English. This one was for the living. He instinctively glanced behind him as he snatched the message into his pocket. He was alone. Still, with an iron-willed idleness, he meandered his way toward the gate, stopping now and then to inspect a gravepost and bow his head in respect.

He did not read the note until he reached the Embassy. When he did, he laughed to himself and placed it in the burnbag. It began with Harper's false name as a salutation. Then it said, "Thursday, 8 P.M., Okura main lobby. Took the back way into the cemetery, over the wall. Easy climb. Sincerely, B." Harper particularly enjoyed the "sincerely."

There would be time later to chide Birch about his methodology. You don't escape notice by hauling yourself over fences in broad daylight. The point was to hide in ordinariness, the more so in a place where your very size and hue singled you out. For the moment, though, Harper's main concern was preparing for the meeting at the Okura. None of this involved Birch directly. The less he knew of the surveillance the less likely he was to give it away by an unconscious glance.

The station chief handed Harper the best specialists who could be assembled on such short notice. Together they planned the net. There was no question of putting a bug at the dinner table. There were too many different restaurants at the hotel, and its staff was generally much too honorable. It was decided not to wire Birch for sound; the others on the team felt that he had not been sufficiently well tested to risk it. Moreover, the station chief felt—as he dutifully recorded in a memorandum to file—that it might eventually become necessary to wire somebody up against Birch and that there was no use placing him on warning how it was done. It rarely hurt a man's career to raise more reservations than, in retrospect, the occasion warranted. So Harper took the station chief's skepticism without undue concern. He did not fear his superiors' caution. What he feared, with good reason based on bad experience, was zeal.

In the end the net was a simple one. Three spotters, all with hidden lenses, were choreographed to watch and shoot for brief periods and then get the hell out. Two were Japanese-Americans—good anywhere in Asia. The third was a Caucasian woman, pretty but not memorable, good all over. Once the pair of marks entered the restaurant, no attempt would be made to observe them until they emerged.

Two autos were stationed to follow them when they departed, and a backup team on foot was ready as well. Harper rode with George Dankers, who had a reputation for rolling surveillances. The sun was still up when they pulled into position, so Dankers stayed well away from the entrance. Harper was on edge, and Dankers' absolute calm only made it worse. His big head relaxed into the cushion of the headrest, revealing a neck aswirl with half-shaved hairs like pinfeathers.

It was getting dark by the time Harper finally saw Birch in the distance walking up the steep hill. Birch glanced in at the Embassy and nodded to the guard at the gate. He did not hesitate when he reached the hotel. He appeared to be remarkably relaxed, considering. In fact, Harper swore he saw Birch whistling.

"The mark?" said Dankers, lazily.

"That's ours. We don't know the other."

"Right. Our guy seems pretty young."

"They all have to start sometime."

Dankers smoked a cigarette, ticking the ashes out the window, until it was time to move up. Then, without a word, he flicked the cigarette away, started the engine, and eased the car into the secondary position where they could see the front doors. In the overhead lights they could make out the woman distinctly when she came outside. She stopped just beyond the glass doors, waved away the majordomo, then opened her purse and began rooting around in it. That was the signal that meant she had made a clear shot, full front, and the pair was in the restaurant. She finally pulled something from the deep leather bag and then quick-stepped her way to the street and down the hill.

"Beautiful," said Harper.

"A little stumpy for my taste," said Dankers. "Cold as hell too. Ought to take lessons from the local talent, I'd say."

Harper sat wrapping and unwrapping a keychain around his fingers as Dankers lighted up again and got comfortable for the long wait. Taxis swarmed in and out of the hotel grounds. Dankers, once his cigarette had burned down to his fingers, seemed to fall asleep. Harper kept watch and tried to pick up random words that floated across the darkness from where the couples paired and separated. But in the end it was Dankers who noticed.

"Here they come," he said.

The starter whined off the stone wall as Birch and the other man appeared in the doorway. Harper would have recognized the other man on the street by Birch's description alone—perhaps even without it. They sent their best and their worst abroad, the worst to keep an eye on the best. Through the field glasses he could see the heavy line of his eyebrows across his broad forehead, the pendulous bags that hung beneath his eyes like the flaps of ornamental skin that make certain lizards too repulsive to be attacked for food. A good rural Party member, Harper guessed, who chanced upon the right patrons and was rewarded for his loyalty and lack of shame by being given a leave of absence from the classless society. It was a long way from the farm, but the Slav had not thrown off the marks of

his past as well as Birch had. The man who called himself Nowicki looked like a person who had known the soil so intimately that he would never be able to wash himself clean of it.

The Slav took Birch firmly by the arm and tugged him in the
direction of the taxi stand. Birch held back. Harper cursed him
under his breath. The Slav screwed up his bushy brow and
buried his chin in the folds of his neck in an exaggerated panto-
mime of disappointment. Birch laughed, pulled away, shook
his head no again. The Slav looped his arm around Birch's
narrow shoulders and jabbed a stubby finger into Birch's
breast. It did not work. Finally, the Slav shrugged and shook
Birch's hand, then climbed into a tiny taxi like a bear at the
circus.

As Dankers eased out a comfortable distance behind,
Harper caught one last look at Birch alone under the brightly
lighted awning. Harper could not bring himself to be angry at
him, any more than he had been able to at the cemetery. There
were always spontaneities even in the most careful designs.
Whatever the Slav had wanted of him, Birch had not been
prepared for it. Harper had not trained him thoroughly
enough.

Dankers stayed with the taxi as it twisted through side streets
and then pulled into the jam on a thoroughfare. The night had
started to drizzle, but the sidewalks still whelmed with people.
Dankers kept one car between him and the cab in the slow,
braiding traffic. At the intersections, when the cars momen-
tarily cleared away at the change of a light, the empty pavement reflected the gay neon in bright, warped swirls. Unem-
barrassed by ambition, the city bragged its commerce like a
medicine show.

Ahead of them a car turned off abruptly, putting them
directly behind the taxi. The Slav rubbed at his neck, and
Harper caught a glimpse in the taxi's mirror of the dark stroke
of his brow.

"He's watching," said Harper.

"We'll relieve his anxieties," said Dankers, and when the
light changed he pulled into the center lane and passed.

"Don't worry," Dankers said. "We're not out of the game.

We'll just give the backup the lead for a while. This is nothing. This is a Sunday spin in the country."

When they were a couple of blocks ahead, Dankers pulled into a parking space and let the car idle with its headlights off. All the intervening streets were one way right, and the taxi hugged the curb. Dankers knew what he was doing.

As the taxi pulled past, it was all Harper could do not to steal a look at the mark. Then the second team came by and the driver gave them a wink. Dankers darted out behind, cutting off a panel truck. Its horn barked.

"Pushy little bastards," said Dankers.

"They've learned from us not to be polite."

"Polite, hell. Two things about these people's anatomy you get to know real well—the tops of their heads when they are bowing and the tips of their elbows when they are pushing you out of the way."

The panel truck had swung out around them and was trying to nudge in behind the backup car. Dankers didn't yield, so the truck sprinted up to ace its way in front. Failing that, too, it cut off the taxi just before the light.

Dankers was ready for it. The car lurched to a skidding stop just as the taxi slammed slow motion into the rear of the truck with the pathetic sound of thin metal giving way.

The backup car missed the collision too, pulling around the accident and through the light. Dankers followed it and stopped at the first available spot.

"It's always some damned thing," he said.

Harper looked back and saw the taxi driver gesturing wildly at the driver of the truck, pointing at the smashed-in front of his cab. Dankers told Harper to face ahead.

"I guess you just averted an international incident," Harper said.

"We want to see what our mark decides to do," said Dankers, adjusting the side mirror with the little handle on the inside. "The backup driver will get pointed in the other direction in case the guy hops another cab going that way. I don't mean to kid you now. We're spread just a little too thin."

"Birch should have stayed with him," said Harper. "Damn him."

"He's out of the car now, and he's pissed. I love to see a Russian pissed."

Dankers blew smoke through his smile, but then the smile went away.

"Shit," he said. "He's on foot now."

They got out of the car just in time to see the Slav disappear around a busy corner.

"Maybe he'll make his meet after all," said Harper.

"If that's what this is all about," said Dankers, breathing hard.

They slowed up when they got within half a block of him. It was one of those honky-tonk neighborhoods with something for everyone. Lust for Daddy, games for the kids, somebody else cooking the meal for Mom. Movie houses bathed in gaudy light, little restaurants, pachinko parlors, massage joints. It had stopped raining, and the crowds spilled into the street, parting as if by some signal to let the occasional car inch through behind them. Mom and Pop and baby-san. Hustlers, barkers, touts. Single men queuing up at the girlie shows. Hawkers selling little mechanized monkeys that slapped their cymbals and snarled.

The Slav pushed his way through the swarm, a head taller than anyone else.

"They keep any of their safe houses in this neighborhood?" asked Harper.

"Beats me. I just drive."

That would not be a bad night's work at all, Harper thought, getting one, maybe two of them IDed, sniffing out one of their private spots, even if Birch wasn't along to give them inside eyes. The Slav stopped in front of a doorway, studied the sign for a moment, then stepped inside.

"There it is," said Harper. "I'll call in some help."

"Can you beat that," Dankers said, laughing, as Harper thought at first, victoriously. "I should have known. They're all alike."

He took Harper by the arm and led him into the street. As they passed opposite the place, Harper saw the dingy little doorway bordered on both sides with faded lavender photos of women, scantily clad.

"There's your safe house," said Dankers. "Magic fingers and mystery."

It was a massage parlor.

8

Harper assumed nothing that night. He called in another team to cover the rear entrance, and when the Slav finally departed nearly two hours later, skew of tie and stumble of step, volunteers went in to check it out. Close investigation revealed it to be, to Harper's chagrin, exactly what it seemed. The girls were clandestine professionals, but not in the collection of secret knowledge. None of the patrons who had been in the establishment with the Slav had anything but their private fancies to hide. And the whole operation came to be known around the Tokyo station as "The Great Debriefing."

But the night was not a failure. The snaps the photo team made in the Okura lobby made it an easy matter to identify the Slav. He was a nominal member of a Soviet trade delegation negotiating a deal for heavy machinery, and we had quite a line on him. His forte was not international finance. As Harper had guessed, he was up from the farmlands where his primary cultivation had been Party discipline. He had taken the name Nowicki when he entered the military intelligence service, GRU. The old one had lost its usefulness: Pavel Nicholevich Rubashov, son of one of the victims of the Stalin show trials.

The central files in Langley had quite a collection on the

elder Rubashov. He was one of the original Bolsheviks. As a young man he had taken on the difficult assignment of organizing the agricultural sector. He had not been terribly successful, but the Party hierarchy had its reasons for not placing the blame on him. Instead, he was credited with a series of grand and fraudulent successes. Nicholai Pavelovich Rubashov lived well, especially after the man to whom he had pledged fealty became leader of the Party. In the files at Langley we have a rare photograph of the elder Rubashov and Stalin himself playing with young Pavel in a dacha outside Moscow. When Stalin brought Nicholai Pavelovich up into the leadership, Rubashov left his wife and son back in the farmlands rather than moving them to Moscow. This was the stuff of workers' myth, and the gesture did not go unnoticed or unexaggerated in the Party organs. Secretly, perhaps, Rubashov wanted to protect his family from the intrigues of court.

As the father continued to weight his lapels with hero's iron, the family led a relatively simple life on the land. The father's success protected son and mother from having to slip into the heresy of private cultivation—though it was this gray market that kept the state from famine. While the other families around them tilled their own little plots of land after putting in time on the communal spreads, the Rubashovs became the very models of Soviet agriculture. And if they took a handsome monthly stipend from Moscow to tide them through their orthodoxy, this was a point not to be mentioned.

Of Nicholai Pavelovich's affairs in Moscow, reports are sketchy. He was not a confidant of Stalin, but no one was. He was hardly a man to oppose the leader. He had been a revolutionary, true, but he was always a soldier. He followed orders, accepted others' conceptions of the new Soviet state. He saw the personification of the future first in the figure of Lenin, who was remote to him, then in Stalin, who was more real. And he was as unlikely to reject their definitions of the right as he was to deny material, orthodox fact.

This was the very reason he did not achieve all that was hoped for in the farm program. He never departed from the literal gospel of the directives, no matter how unreasonable they

became. If the state genetic doctrine said that acquired characteristics were inherited, he would damn sure have the peasants
out there shucking corn on the stalk and would predict with ut-
most certainty that next year's crop would grow unencumbered
by husks.

When it came time to discover traitors in the larger interest
of worker solidarity, Nicholai Pavelovich clicked his teeth at
every new revelation. To do him justice, he was probably not
particularly surprised when suddenly he found the accusations
pointing his way. At Langley, some students of the era among
my colleagues have contended that Nicholai Pavelovich must
have protested his innocence during the months he was held
incommunicado before his trial. They hold that he must have
been worn down by starvation, fatigue, and abuse until he was
willing to say any words that would deliver him from his
suffering, even by death. But this does not account for all the
facts. The films of Nicholai Pavelovich during his celebrated
trial do not show a broken man. They show him portly as ever,
sitting brace erect in the dock, some would say proudly. And
authoritative reports of his testimony about plotting assassination indicate that he expressed himself with a passion and
remorse that even in our far different system of adjudication
would have persuaded a skeptical judge that he was every bit
the malefactor he claimed to be and that his plea of guilty was
not only voluntary but also deeply felt.

We have never explained this phenomenon to our satisfaction. And even to this day there is a muted but persistent
debate over what techniques the Soviets used to get Nicholai
Pavelovich and the others to perjure themselves so effectively.
Some take the mechanistic view—drugs, aversive therapy, that
sort of thing. This faction has certain peripheral motives, since
it is closely linked with the now-discredited Agency human ex-
perimentation program. Another faction contends that Nicholai
Pavelovich was never a loyalist at all and was only confessing
the truth. The consequences of this interpretation are a bit
more obscure, but they have to do with the view that the
Soviet Union is not the political monolith it appears to be but
is rather an unsteady confusion of conflicting interests held in

tenuous check by terror. This is the psychoanalytic theory of U.S.-Soviet relations, and it always leads to recommendations designed to encourage the better elements within the Soviet political elite, to stroke the ego and suppress the terrible red id.

The psychoanalytic theory is often joined to the idea, which is so close to Harper's heart, that the two opposing cultures will rise and come together as they compete. But on the matter of the Stalin show trials, Harper made quite a different point about convergence, one that was less hopeful and much more compelling. He thought that Nicholai Pavelovich had come to believe what he had testified. Harper did not think the Russian had been brainwashed, nor did he take the confession as accurate. Instead, he recognized that a lie is not simply an alternative to the truth. The two can meet, modify one another, become so commingled that it is impossible to differentiate them or to deny that both have an equal purchase on reality.

Nicholai Pavelovich was no plotter, but he must have had his secret reservations about Stalin, tiny interstices of doubt which, when revealed and put into words by his inquisitors, could easily have seemed to a man of such loyalty as apostasy. He had done enough ruthless jobs, not even stopping at murder, that he could hardly deny he was capable of acting upon such malign intentions. He was capable, and he had the hidden purpose. It was a small step from there to the confession. He had lived the deceptions of the state for so long that his final testimony—wired together with the secrets of his hidden, guilty heart—came to be Nicholai Pavelovich's truth.

After his fall, his son and wife were not allowed to move. They kept their simple dwelling, but the subsidies came to an end. They worked on the collective like everyone else, selling home-grown radishes on the gray market now to stay alive. When the war came, young Rubashov fought well in the infantry and was lucky enough to avoid the taint of capture, which he feared more than death. His past was lost in the chaos of the long, Hitler winter. He received a field commission, which he retained at full grade into the armed peace. Though advancement slowed, he began to show an interest in Party affairs and found that despite his name he was not rebuffed.

When it came time to revise history, Rubashov was brought forward to denounce what had been done to his father. He went back to his native town, wrote pamphlets with the strongest permissible wording. He contrasted the enlightened newer order with the reign of darkness that had broken his family. It is open to question whether he felt any vindication, let alone joy. The prose he wrote, even on this painful subject, was as lifeless as the turgid agricultural exhortations his father had written. The younger Rubashov wrote because it was expected of him, just as earlier he had kept his silence. And as a reward for both, he was finally admitted into the privileged orders of espionage and was given an identity without the burden of a past.

The betting in the Tokyo station was that Rubashov/Nowicki amounted to very little. As often as any other, the Soviet society visited upon the sons the sins of the father; the Slav would not be trusted with significant work. Low-level intelligence jobs abroad were a combination of honor and painless exile, like American ambassadorships. For a time, everyone in the station save Harper lost interest in the case.

But Harper moved doggedly along. If the other side at the outset thought little enough of Birch's potential to trust him to a GRU functionary, that was not surprising. The Agency itself would not have sought out a man like Birch with any enthusiasm either. But Harper knew from personal contact that it was a mistake to underestimate Birch, and he fully expected that the Soviets would eventually be brought around.

On a hunch, Harper chose a special safe house for their next meeting. It was in the Shibuya district, snug above a bar on an obscure back street. It could be reached by any one of a half dozen routes, alleywise or by a door facing front. The Agency realtors kept the apartment simple; sleeping rolls and a few low chairs made it look to all the world like just another cramped family flat. The very economy of decoration also made it difficult to bug and easy to sweep. This was an advantage because it was often left vacant and unguarded.

Harper arrived early with the technicians. The room was musty. After the electronics men had swept every corner and cabinet with their magic wands and declared it secure, he opened a window and let in the warm, polluted breeze. Downstairs the bar was open; it was always open. From the window he watched a gaggle of sailors stagger in. The music that rose up from below was country and western, loud and sullen.

At the appointed time, Birch strode straightway up the street, hands jammed into the pockets of his khakis, easy as you please, checking out the windows of the novelty shops. The uniform was an inspiration. A wonderful cover. He did not glance behind him, and from Harper's vantage it was clear that he did not need to. They still were not bothering to follow him.

When Harper opened the door for him, a piercing guitar and whining voice drowned out the forced laughter of the bar girls.

"Kind of reminds you of the International Barn Dance on the radio when you were a kid," said Birch.

"What's an international barn?" joked Harper.

"You know, I never thought about it until now."

As soon as they sat down, Harper brought up the dead drop. He was as stern as he could manage about it. There was a lesson to be taught.

"You aren't supposed to attract attention," he said.

"It's all right," Birch said. "I had checked it out before I made my move. I knew the lay of the land."

"That was another mistake. Never go anywhere near the drop site until you actually make the drop. Now that they have their lines on you, they may start watching. And if you snoop around the same obscure corner twice, they will know something is up. You must always play it exactly as I tell you."

"I'm sorry I made you angry," said Birch.

"The point is, you have to take my instructions on faith. Don't second-guess me. You don't know enough to try."

"Yes, sir," he said with a private's questionable contrition.

The relationship between a handler and his asset is a delicate one, and Harper knew the importance of balancing discipline with encouragement. Having gotten the hard part out of the way, he told Birch that the errors were not fatal this time and that the mission had gone off without a flaw. They had identified their man and, as expected, he was Russian.

"He did just what you said he would, Mr. Harper," said Birch. "He wanted photocopies, but he took the handwritten notes. I told him about the safe and the security devices. He ate it up."

"I'm surprised he didn't ask you to sign the document."

"Why would he do that?"

"He wants to compromise you, and you're giving him everything he wants."

"If I'm going along with him, why would he want to be able to blackmail me?"

"Their intelligence agencies are not paternalistic enterprises. They assume that everyone is in it for himself. They are like nations. They see security only in overkill. They are only comfortable with those whom they have the ability to destroy."

"I hope I didn't screw up," said Birch.

"You played it just right," said Harper. "Don't worry about incrimination. We are the only ones who can put you away."

Birch seemed relieved, and Harper did not feel obliged to warn him that the Agency operated no differently.

"He told me his company would be pleased with my progress," said Birch. "He said my stock would be very high."

"A regular capitalist."

"He gave me another assignment. Here, I wrote down the documents he wanted. I told him I wasn't really sure."

"Is that why you balked at the door?"

"You know about that?"

"He wanted you to do something and you refused."

"Yes, sir," Birch said, and in the long, reluctant pause they heard someone moving on the stairs. A woman giggled.

"What's going on?" Birch whispered.

"Don't worry. They aren't interested in us."

"Who are they?"

"The girls downstairs use the rooms next door when they find a man who's in a hurry."

"Funny place for you to put your office," said Birch.

"Very upsetting, this business downstairs," said Harper. "Used to be a sushi joint. Then the war, all those kids on R and R. Girls all had their eyes redone round. Pumped up their fronts too, some of them. It destroys the real beauty, don't you think?"

A door slammed and the hallway was quiet again.

"Shall we go on?" asked Harper.

"Sure," said Birch. "Sure. Sometimes I get a little confused is all."

"How much did he pay you?"

Birch reached into his pants pocket and drew out an airmail envelope. He handled it by one small corner, holding it out away from him as if he were afraid it was going to drip.

"I wasn't sure whether you'd want to try for fingerprints," he said.

Harper ripped it open carelessly and counted the fresh twenties and fifties.

"There's five hundred dollars here," he said. "You've been overpaid." He was surprised and pleased at the amount. The Soviets were not free with their money, and they didn't let minor figures like Rubashov authorize such large payments. Already somebody else had gotten interested in Birch, and Harper assumed it meant the KGB.

"Where did the Slav want you to go after dinner?" he asked.

Birch looked nervously at the little cassette tape machine that hissed softly at the silence. Then he studied his reflection in the distorting spit shine of his shoes.

"It didn't have anything to do with business," he said.

"Everything is business, I'm afraid."

Birch picked up the recorder and held it flat in his palm. "Do we have to have this on?"

"It makes you uncomfortable?"

"No," he said quickly. "Yes it does, sometimes."

"I'd be glad to turn it off," said Harper, "but why make it seem as though there is something to hide?"

"Leave it on," said Birch. "Leave the damned thing on."

"Have you ever noticed that when something upsets you, you tend to repeat yourself?"

"No," he said, stopping himself and then carefully choosing the next word, "I haven't noticed that at all."

"It's a habit we'll have to work on as we go along."

"I just didn't expect anything like this to come up," he said. "Nowicki seemed like such a, well, such a plain man, so courteous with Donna and all. I was surprised at what he wanted. I didn't know what to make of him or what made him think that I would . . . I certainly didn't say or do anything that would make him . . ."

A loud pounding fist on the door cut him off.

"Snake! You in there? Open up. Time for sloppy seconds."

The banging grew louder, and then it sounded as if the oaf were putting his shoulder to it.

"He won't get through," Harper whispered.

The door shuddered with a heavy blow, but held.

"Holy shit!" cried the oaf. "I think I busted my shoulder."

"Randy," said another voice, "you simple sonofabitch. You got the wrong door."

"Landy-san," said a tiny voice, "is OK?"

"Nothing you can't fix, Dolly."

Steps moved down the hallway. A door shut, and it was quiet again except for the cloying whine of the juke downstairs. Birch still gazed at the door as if the door were a man and the man had a gun.

"Take it easy," said Harper. "They're harmless."

"You set me up," said Birch. "You did this to me. I thought I was supposed to be able to trust you."

"Nobody asked you to go next door and take third place in line. And if you went out on the town a few times with the Russian, that's not going to shock me."

"That isn't it at all," said Birch. "The other night at the Okura was the first time he even asked. I told him I wouldn't go. He said he knew a place where they did things my wife never even dreamed of. I wanted to deck him."

"Dutiful husband."

"What's wrong with that?"

"It's admirable, Birch, and all too rare."

"You've got to understand how I feel about it. I just couldn't hurt Donna that way."

"Everyone has certain limits. We try to respect them whenever we can."

"I appreciate that, Mr. Harper. I really do. I couldn't play along with him on this. Never."

"Nothing was lost this time," said Harper.

"It felt so strange when I came home to her that night. She was waiting for me, full of questions. How did your business go? Was dinner good? Where did we go? Did he ask about her? I felt so ashamed."

"No reason to. I hadn't briefed you how to handle that kind of situation. . . . Now, the next step is for me to get you the material he wants. We can wait a few days to pass it, but in the meantime I want you to buy some things. Nice things for your wife. Keep the receipts and we'll reimburse. But tell your friend you bought the stuff on time, piddled the cash away. Let him think you're in hock."

"I guess in a way I am."

"The more the better."

"I owe you too," said Birch.

"You're a soldier," said Harper, because he could not let his man believe the relationship was the same in both directions. "You're used to having government issue hanging in your closet."

"They don't issue fur coats."

"Only to the most elite troopers, Birch. If you can get him to see some of your new things, maybe your wife's fancy clothes, that will help."

"It's going to make her nervous," said Birch. "She's already beginning to wonder what all the money is for. She's the kind of woman who shows it."

"Good," said Harper. "If she's a little on edge, it won't hurt. She believes in you, doesn't she? Anyway, she'll have the things."

"I hate to lie to her."

"It won't be lying, just sparing her some of your anxieties. Tell your friend she wants to see him again. Yes, that's good.

She was charmed by him and has been jealous that you two have been going nice places without her."

"She never complains."

"You have a good woman there," said Harper. "A real asset."

Harper offered Birch a drink when their work was over, a toast to the continued befuddlement of their adversary. He said the words, sipped from his glass, then went to the window to watch the gaudy dusk descending. The neon was coming up all over the city. Beyond the far rooftops and the flickering marquee of a movie house he could see the graceful, atavistic curves of the neighborhood shrine. It seemed out of place amid all the glass, steel, and poured concrete. They wore our clothes, played our sports, listened to our music, built our products. We bought their goods, dabbled in their religions, parodied their lusty indulgence. Their political system was written on our texts. Our commerce studied their success. Below the window on the street an old woman in traditional robes scraped along on wooden sandals, thin white anklets hiding her skin, a portable radio held tight against her ear.

"They really could be so different from us," said Harper. "Once they were. Do you suppose all enemies end up converging?"

Birch seemed surprised by the word Harper used, but he did not reply.

9

It was at about this time at Langley that the muttering began about whether Harper's personal life was getting in the way of his work. He had just made his proposal to Janet, and everyone knew about it—word for word—because he had used the leased-line telex. Later she joked about how sensitive he had been not to send it enciphered. In fact, the telex was a last resort. She was scheduled to depart Saigon the next day and they were to meet in Bali for a last vacation on her way home to the States. But the Birch case was keeping him in Tokyo. He had to reach her quickly, and the monsoons had scrambled the PTT phone system in Vietnam beyond the hope of recovery. He did not want to lose her.

My colleagues often accuse me of being contrary. They say it is characteristic of my branch to see everything as its opposite, and I suppose they have a point. Nevertheless, I was genuinely delighted to learn of Harper's decision to marry. I have always believed that it is impossible to separate our work from our lives; the word *intelligence* itself implies the intimacy. I was encouraged by his choice of Janet. She had a perfect security profile; no problem there. She was the kind of woman one would expect to be attracted to Harper: the right schools, the

proper family background, the shared graces that help us per-
petuate our class. I was quite confident that they would settle
into a comfortable situation, and the fact that she had seen him
at his worst made this all the more encouraging. The marriage,
for both of them, was a kind of partnership of us against them.
This is an important orientation not only for couples and
classes, but for nations as well.

There were, however, some revealing materials that showed
up in Harper's files dating from about this time. I discovered
them much later. They were like faded photos in an album,
keepsakes, keys to sentiment. One was a piece of footage taken
one summer evening from the surveillance van Harper had for
a time posted on Birch's street. Back in those days, of course,
we were not troubled by lawyers' niceties in our operations.
The Bill of Rights, we were happy to assume, was territorial. It
did not apply in hostile jurisdictions. Now this assumption has
been very nearly reversed by those whose ideas of propriety
are uncorrupted by experience.

The film from Harper's safe showed Birch and his wife coming out of the house with a wicker basket and a blanket. Birch
pauses at the doorway, apparently running down a mental list
to make sure nothing has been forgotten. The film jostles at
this point. I suppose the cameraman was bracing for the chase.
But the Birches weren't going far. They spread the blanket on
the parched grass of their front lawn, which is so small and cut
up by walks that the cloth laps over the concrete on two sides.
As Donna unpacks the food and plates, her husband takes out
two candles, sets them symmetrically on the ground, and with
some difficulty from the dry, fickle breeze, lights them with a
match. Jeeps jerk past, deuce and a half trucks from the
nearby post. The Japanese man who shared their duplex can
be seen peeking out the front window and shaking his head at
this strange rite. Later, when it gets dusky, Birch is bold
enough to give his wife a little kiss. She accepts it, pats his
cheek, but keeps him at a distance and looks about to see if
anyone has witnessed their indiscretion.

Another memento is a tape cassette, erased but for one small
piece of conversation, the kind that would not find its way into

an official transcript. It must have been picked up on a mike in their dining room. Internal evidence suggests that the two of them were waiting for a visit from the Slav.

"Where are you, love?" Birch calls.

". . . kitchen."

"I'll do some of that for you. We're running a little late. We don't want to be in a mess when he gets here."

"I'm sorry. I got behind."

"No problem. It isn't important."

"Here, you take this and I'll just put this over here. It'll be faster if I just hand these things to you . . . Did you see in the paper about that airplane crash?"

"Terrible," he says.

"I don't like you to fly."

"Only when I have to."

"I want to be with you when you do."

"I'm not going anywhere," he says, laughing.

"When you do."

"Sure."

"I want to share."

"I think we're pretty safe, don't you?"

"I'm being silly," she says.

"Silly nice."

I suppose it was not entirely healthy that Harper was looking to his asset as an example for his own life. The bonding that resulted was ultimately quite difficult for Harper to break. But let my colleagues say what they will, sentiment itself is not always a liability in an outside man. It is a question of degree; to manipulate, you see, one first has to have the sense of touch.

Meantime, at Langley the stakes in the Birch operation were rising steeply. We had had our grim meeting on Black Body, and we were hungry for options. The situation was getting worse by the day. Our counterintelligence teams had less and less to report. While it has become commonplace among the dreary optimists to assume that increased communications will bring the Soviets and the United States together in greater accommodation and safety, the reality is that our greatest safety is an intercepting the very communications they most want to deny us. Black Body cut us out, and the atmosphere in the upper levels at Langley was near panic.

At this point, of course, the Tokyo station was not consulted. The matter was too sensitive for foreign dissemination, and the discussions were all in house. Several alternatives were suggested, all of them designed to trick the Soviets into abandoning their communication security success. And the only thing that would lead them to do that would be a belief that we had already solved the Black Body riddle and were reading their messages in the clear.

One proposal involved a series of arrests of known deep-cover agents in the hope that one or more of them had been fitted out with the system. Even if we did not manage to get our hands on the equipment, so the reasoning went, we would create the impression—or at least the doubt—that Black Body had been compromised. This approach had severe limitations. One was the old truism about the devil you know. Once an illegal agent has been identified, we like to keep him in place because by surveilling known illegals we can get information about the parts of the network that are still hidden. How do they operate? What are they after? How active are they? To make enough arrests to have any chance of worrying the Soviets about Black Body's integrity, we would have run the risk of leaving ourselves even more deaf. Moreover, the illegals we had identified had all made their infiltrations before the deployment of Black Body. The newer illegals, the ones born to the innovation, were all unknown to us. This point would not have been lost on the Soviets as they tried to evaluate the meaning of our offensive. The ruse was too thin. Some numbers cruncher calculated the odds of success, and they were very discouraging. The Sisters, who had responsibility for counterintelligence in the United States, were sharply opposed. They had to have somebody to surveil.

A second approach envisioned an elaborate stratagem in which we would pass to a suspected illegal certain key items of information and hope that he would report them via Black Body. Then we would let it be known, by actions characteristic of a damage containment operation, that we knew the information had been reported. This approach, too, had its drawbacks, not the least of which was the sheer guesswork of the thing. It depended on our hunch about who might be a new illegal and on a precisely choreographed series of actions that is virtually impossible to bring off in this world of blindman's buff. One slip—one failure of an illegal to report, even an unforeseen delay—and the game would be lost. Even if it did work perfectly, it would not be guaranteed to focus doubt on Black Body. As the Germans' unwillingness to abandon Enigma during World War II indicates, an intelligence service will always look first to human rather than technological weakness when it is trying to locate a leak.

The third proposal was marked by directness and synthesis. One highly controlled operation of the sort suggested in the second alternative would sow the doubt. Then we would have one of our own double agents give his Soviet handlers a hint that we had been reading Black Body all along. What we needed was a good contact and a straight shot to the top of their espionage hierarchy. It was unanimous that the man who called himself Nowicki was just a beadle and would not be trusted by his superiors in such a grave matter. We had to get to a higher echelon. Like anything else, the value placed on a piece of intelligence is not entirely a matter of merit. Its reception depends on who sponsors it. In a bureaucracy, even knowledge is politics; the disinformation we had in mind needed a powerful voice.

Apart from the fact that he had not yet been pried away from the Slav, Birch was very well positioned for our purposes. He had an electronics background, and since we knew that the Soviets were using Black Body in Asia too, it was natural that Birch might come into contact with any method we developed to penetrate it. The fact that the Soviets had approached him first added somewhat to his credibility.

The question was how to get the Slav replaced by someone in the KGB, someone impressive. The amounts Birch was being paid certainly indicated state security organization attention. But for whatever reason, the top people were holding back. At Langley, based on a detailed review of the record, we tentatively decided that Birch still seemed too independent. The KGB did not feel secure in its control. Like Harper, the Soviets recognized Birch's inner resources, but unlike Harper, they were wary. The Russians are quite conservative. Before they make a move they like to make sure they have their man on a string. They want to see him dance. It is the only kind of authority their culture understands. If Birch was to be of use against Black Body, it was decided, his submission to their wishes had to be made to appear absolute.

Harper's likely reluctance to take the necessary direction from above was duly noted. But an agent was an agent, and his personal problems could not be allowed to interfere with the larger design. His catastrophe in Vietnam notwithstanding, he would have to be brought into line. The orders were sent.

10

A picture of the President was the only decoration on the station chief's wall save a handsome piece of calligraphy. Rumor had it that he had gotten the latter from one of the State Department's China prophets before their fall. It had hung in all the station chief's many offices since then, even during the bad times when the man now pictured on the opposite wall was still a congressman rising on the broken backs of our Asian foreign service. Now, of course, displaying the calligraphy was no longer controversial. The disgraced have all been rehabilitated, even those who probably had deserved to fall. And the station chief's record remained untainted. He had survived the period that had destroyed his friends, and he would survive all the others, whatever their orthodoxy, because when he had a thing worth showing he simply tacked it up on display and said nothing more.

"Come in and sit down, Harper," the station chief said. "I'm sorry to keep you so late. Will you have your whiskey neat? I'm afraid I forgot to have the girl bring ice."

"Plain soda for me," said Harper. "I have some more work to do after."

Perhaps it was not exactly that the man refused to talk about what he displayed. There were those who claimed that the calligraphy's origin was common knowledge simply because the station chief had always made sure that it was. But if that were so, he had done it indirectly, which was another technique of survival.

"A Mao Tse-tung poem," said Harper.

"Is it really?" said the chief. "I didn't know you were a China scholar."

"Someone told me once."

"Uhm. It's something about nature, I think, or guns. Maybe both. Never did master the language, you know. Great failing on my part. Brain just doesn't have the connections for it, I guess. If you don't mind, I think I'll pour myself a brace anyway."

"Please do," said Harper. "It would make me groggy at this point."

"Just the thing. It's properly late in the day, don't you think?"

"I want to get through the surveillance logs."

"Keeping him on a very short string, are you? I don't want to discourage you from taking pains, but simply surveilling Birch isn't going to advance the case very much."

"I'm a patient man," said Harper.

"You have never gone up against the Russians before."

"Only their friends."

"This is not the same. You are not working with Asian middlemen this time. This is a European maneuver, classic as Clausewitz in its way. The Russians know how to exploit an enemy's patience."

He carried his drink in his thin, veiny hand and sat opposite Harper on the couch. One hand occupied, he hitched the other by its thumb under his belt as if to restrain himself from any spontaneous, revealing gesture.

"Let me put it bluntly," the station chief said. "I leave diplomacy to the foreign service. My operations are pure. I do not have grand designs."

"It is getting a little late," said Harper.

"Sit down. Be a little more clever with me. You know that something has come up, but leaving me sitting here alone with my whiskey is not going to make it go away. I thought perhaps your confidence had returned to you by now."

"What makes you think I lost it?"

"I read your reports. You did not exactly write an indictment of Bartlow. You were self-critical. You did not go after him at all."

"That took care of itself," said Harper.

"And you survived."

"It is something I would have assumed that you would appreciate."

"I realized how dangerous you could be," said the station chief, "if that's what you mean."

"Did you ever read Dryden's definition of satire?" said Harper.

"I'm afraid not," said the chief. "It's as foreign to me as the ideographs on the wall. I read fiction, for the technique, but never poetry. I studied medieval history. Good preparation, as it turned out. Even a king has to beware of his barons."

"And the barons, their king," said Harper. "I've been reading some history myself."

"Well you might. One of Bartlow's problems was that he never had. He believed all that demagoguery about history repeating itself. He was trying to end World War II again. And he was going at it so hard that he didn't realize he wasn't dealing with the German general staff. History isn't the same damned thing over and over. It is a succession of different things, and the only thing they have in common is being damned."

"I don't have any opinions on historiography," said Harper.

"Bartlow had his strengths," the chief went on. "I've watched him work for a long time, suffered him. He was in China too, as I suppose you have read. But he was always a man who knew his audience. He reported what they wanted, one hundred percent. Whatever else that can be said about what happened to your network in Vietnam, it was not destroyed because Bartlow did not understand what was expected of him."

"You are saying that I failed to protect them," said Harper.

"I am sure that the thought has had its day in your reflections. Here, let me get you some scotch now. I believe you might be wanting it—if only to splash in my face. . . . We never did teach them how to make liquor properly. Their Suntory all tastes the same to my palate. This is ours."

"We can't make scotch either," said Harper.

"But we've had the good sense to import it." The chief handed Harper the glass and returned to his perch in the opposite corner of the couch. "Langley has reasons to want your operation to push ahead quickly. They are pressing for action. They want to draw in the KGB."

"I suppose they know how that can be done," said Harper.

"Yes, in fact they think they do."

"You'll excuse me if I feel a bit of *déjà vu*. I must have a reputation as a man who will sacrifice—others."

"I knew you would be wary of this, Richard. But it is nothing like the last time. We are not putting Birch in jeopardy. Don't be fooled by *déjà vu*. They say it is a momentary slowing of the mind."

"I've brought this man along at the pace I have seen fit. He has made no missteps. I cannot be blamed for the Soviets' indifference."

"But, of course, you can be."

"Perhaps someone else would do better."

"There is no way of knowing that," said the chief. "If our mission is to succeed, Birch will need a lot of work. You have his trust, Richard."

"And now I must abuse it."

"You must persuade him to do what will probably be anathema to him," said the chief. "It is not a duty anyone would enjoy. But you will not be asked to risk his injury. There is no hidden menace here. At Langley they think Birch must put himself more in the Soviets' control. They don't think money is enough. Ideology would not be credible. They want him compromised sexually."

"He will not do it," said Harper. "I know the man."

"It won't be easy to convince him of the need. I think I have gotten the sense of him from your memoranda. But the judgment is that you must succeed."

"Who are the people who plan these tawdry things? Schoolboys peeking into the ladies' room?"

"In this case, I concur with them, Richard."

"I will not be responsible for what happens," said Harper.

"Unless you fail."

They say that depression is a chemical event. The gray, choking substance that leaches out the spirit is said to result from some electrolytic imbalance in the brain. But if this is so, Harper thought, the poison crystals are formed on the template of experience. And like flakes of snow, no two of them were alike. Each depression had its distinctive qualities. Harper's had the bitter, raw smell of Saigon at night.

Janet was already in bed by the time he got home that evening from his meeting with the station chief. He leaned over the light bedclothes and kissed her cheek. She stirred, half awake.

"You're late tonight," she murmured.

"I had to attend a history lecture."

"Mmmmmm. Come to bed soon," she said and sank back into sleep.

He stood there for a moment in the dim light of the doorway, looking down on her, easy and secure in his bed, and shuddered with disgust. How could any man make a promise of faith in this world? He closed the door and went to the living room where he poured himself a drink.

The voices on the radio were too loud when he switched it on, so he turned it down to a whisper. A small breeze rang the wind chime outside the open window. The talk was too low to be recognized. It was only rhythms and inflection, like Vietnamese. In the courtyard a rope banged up against a hollow pole. Pok-pokpok-pok-pokpok. The night rustled with insects. In the dry season the air after curfew was cool and the sky empty and deep. Pok-pokpok-pok-pokpok. All that broke the silence was the sound of distant conversations and the rhythm of the night street vendor's clapper, hollow and lonely, like receding footsteps in an empty room. Occasionally a rifle discharged, but always far away. Saigon at night was war at one remove.

But war was different in the daylight. It roared down the street in a big deuce and a half loaded with explosives and steel. It clattered overhead, a helicopter bringing home the dead. It streamed into the city in long, pathetic lines of refugees.

Harper, in his tropical cotton shirt and trousers, stopped every morning at the Caravelle for breakfast with the correspondents up early for flights to the field, trading information, watching the first air strikes of the day blooming on the wide plains around the city. At first his work was purely analytical, as it had always been at his other postings. But it was different here, and he found it, despite himself, exhilarating. Here ideas translated directly into action. Reports were not just documents to circulate; his words were deeds. If he managed to trick out of the clutter of data the presence of an enemy regiment where it was not expected, this could set into motion a flock of choppers on the assault. Men went into battle depending upon Harper's wits. He was engaged and alive.

So when Bartlow, the station chief, gave him the chance to get one step closer to the action, he seized on it. Bartlow wanted Harper to take over a network of agents from Tom Johannes, who was being reassigned to Langley. Johannes was one of the Founding Fathers, an agent who came over to the Agency from the OSS. Like many of the old-timers, he was brilliant but independent. He had been known to drop out of contact with his station for weeks, only to surface with some astonishing bit of hard intelligence and very little to say about how he got it. When Harper said taking over for such a man was a great honor, Bartlow dismissed the thought.

"He's getting old. Won't move. I had to cut him."

"You're the man who decided to take him out of the field?"

"He's done absolutely nothing with the net," said Bartlow. "He's lost his nerve."

But when Harper met Johannes to prepare for the transition, Bartlow's description seemed all wrong. The years showed on Johannes. But dressed in a polo shirt and knife-edged white trousers, he looked ready to whip you at tennis at the *Cercle Sportif* or to slip a stiletto into your back without scraping a rib. He was still trim and agile. The lines on his face were long and jagged, like all the borders he had violated. He wore a map of the world in his expressions, and it was definitely not a place for people who had lost their nerve.

"The key man is Do Van Tri," he said. "I'll make the formal introduction. He's Chinese, a merchant dealing in anything that carries a price. He has business with both sides, and as far as he's concerned, loyalty is just a matter of rates of exchange. You can trust him the way you can trust the gnomes of Zurich. He'll always go with the highest bid.

"The others," he said, fanning several photographs in his palm like playing cards, "they do what he tells them to. Each has his own contact with the Front, and each is out for himself. But they are all afraid of Tri—and for good reason."

He went down the list: hustlers, black marketeers, pimps, smugglers, dealers in dope. He dealt the photos out on the table one by one. Some had histories that dated back to the Binh Xuyen. But that era was over. Crime was no longer its own sovereign state. The conspiracies had become loose and shifting. They inhabited the no-man's-land between the warring armies, and it suited both sides that they should.

"You ever play one of these hands before?" Johannes asked.

"I've been an inside man."

"Bartlow isn't looking for intelligence out of this. That's something you ought to know going in. What you get from these guys isn't worth a damn. They could get you something good if they wanted, but they aren't stupid. They like to keep both sides in balance. They like to stay alive. Bartlow knows that, but he thinks these hoodlums are the key to settling the sonofabitch. He doesn't want information. He wants action. When the time comes, he thinks they will broker a deal."

A helicopter passed low over the building.

"You think he's a fool," said Harper over the thudding rotors.

"He has dreams of glory," said Johannes when the chopper had receded into the distance. "You've got to watch out for a man with dreams of glory."

"They say that maybe you haven't made the most of these assets," said Harper.

Johannes pulled on the cigarette, then held the smoke in his open mouth, like sweet cream, before gulping it down.

"For more than a year Bartlow has been pushing me to try to use Tri as a lever. He thinks he can pry apart both sides and come up with a new coalition. Harper, I don't know when it was that the Agency came under the control of people like Bartlow who think you can change history with a few gimmicks and a lot of money. The damned Brits started it, I suppose, during the War. All that fancy gimmickry before D-Day, and then they missed the real chance with the German general staff. We rode them pretty hard about it, after the whole thing was over. I guess if you're young enough you might take the second-guessing seriously. It's always easy to imagine the shots we didn't play dropping straight into the pocket. Ugly little facts don't get in the way. But to me, history is a juggernaut and the best you can expect is to catch a glimpse of it thundering by."

"We've had some successes," said Harper.

"We have. We have," said Johannes, and for an instant Harper saw the fatigue behind the mask. But then Johannes' eyes focused again, not so much to see but rather, like a glass in the sun, to burn. "We've had our way from time to time. Iran, Indonesia, Guatemala. But for how long? Do you think it matters so much whose bastard is on the throne? Don't forget that we had our way here too. That little sonofabitch in the palace down the street is our sonofabitch, and I don't see that it has made a hell of a lot of difference."

"You have to make the effort, don't you?" said Harper.

Johannes stubbed out his cigarette in a brass ashtray cut from the base of an old artillery shell that would no longer fit the modern bore. "I think you're exactly what Bartlow is looking for," he said.

There were no more insults. Johannes made the appropriate introductions, gave the standard briefings. Then he filed a sarcastic cable to Langley (which Bartlow forwarded because he thought it proved his point) and then resigned. Johannes left Vietnam but stayed in Asia. And when things fell apart, as he had predicted, I took it upon myself to fly out and try to persuade him to come back to help pick up the pieces. He told me to fuck off, and I admired him for it.

He was right about Harper being what Bartlow wanted, but not because Harper was naïve or enthusiastic about covert actions. Harper was vulnerable to this kind of plan because it seemed to him to be a form of convergence. The two sides would be brought together on neutral ground. Security and stability were to be found there. It was so much readier and less violent than conquest. Bartlow had found Harper's greatest weakness and infiltrated there; Harper was vulnerable to hope.

But he was not an ingenue. He never trusted Do Van Tri or his henchmen. He despised their perfumed insouciance, their golden grins. He thought little of the elaborate precautions they took in setting up the meetings. It was an open secret what they did, and Harper suspected the whole melodrama was designed only to increase Tri's price. Still, Harper went along, never contacting Tri at his business office in Cholon, adhering to the furtive rituals and treating the man with a respect he in no way felt.

Under Bartlow's prodding, he often broached the subject of Tri's acting as an intermediary with the people in the Front from whom he purported to obtain the information he was selling. Harper also contacted others in the cabal—particularly Tran Van Binh who operated in the Delta—with the same pitch. It was never flatly turned down, though it never met with enthusiasm either. Bartlow's plan was not precise, but the station chief insisted that preparations be made against the day when an internal political settlement could be consummated. To that end, he cultivated the silent dissidents within the government and urged Harper to keep the issue alive with Tri, however little demonstrable progress there seemed to be.

It is one of our sources of pride at Langley that we always keep an open mind. No alternative is ever dismissed out of hand. Imagination needs running room, and we try hard never to stifle creative thought. This is why during the purges it may have seemed that we lacked all restraint. Nothing was unthinkable to us. Even a far-fetched scheme like Bartlow's was not brought fully to a halt. All we warned about was discretion and making commitments—the one because intrigue has political risks if it comes to light, the other because we reserve for ourselves the ultimate decision of peace. But it is fair to say that no one at Langley held out much promise for Bartlow's plan. In this case, the only fair criticism of the apparatus is that it tolerated a fool's fantasies.

When the danger finally came, it was Tri himself who announced it. As usual, a boy showed up at Harper's door with a bundle of joss sticks. Harper offered to pay, and the boy ran away. It was the signal that Tri wanted a meeting, and the boy's improbable self-discipline was a symbol of Tri's cruelty and strength.

That evening after dusk, Harper wearily hailed a taxi on the rain-drenched street and rode it through the monsoon to the refugee quarter. The rains rolled off the sea in great waves. It was the season when all the illnesses of the spirit seemed to hang heavy and wet in the air. At the spot where the railroad tracks crossed Cong Ly Street, he had the driver pull over and paid him. Then he made his way into the cramped, muddy alleys between the hovels. The rain beat on the tin snares of the roofs. Here and there the yellow light of lanterns silhouetted a figure watching him from a doorway. The police did not venture into these precincts often. Westerners were never seen, except for the deserters, who tried very hard to hide. Here lawlessness was even greater than war. People died unnoticed, and no one bothered to count the bodies.

Tropical lightning burst in the sky like a flare. The defenders of a nearby bridge fired into the black waters to ward off sappers. The thunder rolled in like bombs. And it did not take Harper long in the shadowy and hostile labyrinth to lose all sense of direction. In the past, a man had always come to guide him, but as he continued on deeper into the darkness, he wondered whether this time he might not be truly alone. He stopped and turned behind him to get his bearings. It did no good. When he turned back he discovered a man standing there like a phantom. At first all he saw was the gun. He froze. He was trapped. Then the lightning came again, and it glinted off the familiar gold talisman in the man's hand. It was the medallion Tri wore around his neck, their sign. And so, to the footman from hell, Harper whispered, "Thank God," and then he followed.

"Mista Ha-pa," said Tri's voice as Harper entered the hootch. Tri sat at a small table flanked by two men with rifles ready. He was throned like a warlord. Rain dropped through the roof and splashed against the edge of the table.

"You have tea?" he said. Harper nodded. Tri wore a camouflaged uniform of some sort, tailored, pressed, neither theirs nor ours. His pockmarked face shone with perspiration in the jaundiced light of the lantern. His heavy lips parted, revealing the gold.

"Of course," Harper said.

Tri clucked out something in Vietnamese. An old woman emerged from a dark corner, set two places, and poured the steamy liquid over the crystalline sugar in the bottom of the cups. When Harper took a sip, the sweetness coated his mouth like sickness.

"You have learned something?" he said.

"Di di mau," Tri said sharply and the woman disappeared through the canvas flap of the door into the rain. "She has kindly offered us the use of her simple home. We return the courtesy, in dollars MPC."

"She can be trusted?"

"As long as she hears nothing."

"Now to business," said Harper. "The usual terms." This was the way it always was handled, Harper bidding the common figure, Tri insisting the matter was special and thus required further negotiation. Tri respected the usual terms and always accepted them in the end, but he felt obliged to test them each time. Maybe green money would be more appropriate? Maybe gold. Piasters, Harper would say. The American, he does not understand that he is at our mercy. Ah, the American. He does not yet appreciate Asia. But we will accommodate him. We will do business his way because he is our honored guest.

"No rates tonight, Mista Ha-pa," said Tri. "This is a different kind of palaver."

"You know that I am not free to change the terms of our contract."

"The contract has become a problem for me," said Tri. "I am afraid that we must call an end."

"What do you have for me on such a night that is worth so much?"

"Maybe in time you will think this was a very, very good night," said Tri. "Maybe you will look back and wish you were wet again—and safe."

"The rainy season ends and the fighting begins. I know the forecast."

"This time the danger is to us. I give you this information free. No pay. It is why this must be our last meeting."

"I don't understand," said Harper, pulling the cloth sack with the money out of his pocket and swinging it in front of him. "I am prepared to continue our business."

"We hear things, Mista Ha-pa. We hear that your Mista Bartlow makes promises about the future of our business that are not necessarily right. We hear this, and so we must assume that others hear it as well. They will not like such news. They will think we have gone beyond our tradition. They will fear that we have certain designs on their power. This is not so, as you know. We are simple men. We are merchants. What do we know of war and peace? All we know is how to protect our interests. But your Mista Bartlow is putting this in danger."

"I'm sure there has been a misunderstanding," said Harper.

"We do not blame you. Do not be afraid, Mista Ha-pa. You are safe with us because we know this man Bartlow is not within your control. But we cannot afford to be in opposition, you see. We yield to all authority. We swim with the tides. We know what your man has said about us, and so we must break off contact."

"There is going to be trouble?"

"We are taking the steps to avoid it," said Tri.

"But the others, are they in jeopardy? Who is threatening? The government? The Front? What have you heard?"

Tri sat back and turned the circle of gold on his finger. It was not a token of faith. And he had no intention of answering.

"We have heard enough to beware," said Tri.

"When will the trouble come?"

"Enough," said Tri. "The advantage of giving you this information free, Mista Ha-pa, is that we can choose how much to give. Please, finish your tea and my men will see you safely on your way."

There was no arguing with him, and the next day when Harper tried to make contact again there was silence. The boy with the joss sticks never came.

Bartlow was furious. He blamed Harper for not having established a close enough relationship with the thieves. Tri's expression of concern about what Bartlow had been saying around town did not impress him. He said it was only an excuse.

"Cold feet, Harper," he said. "I've seen it before. They have a case of the jittery asshole. Something is up, and they are worried about it. Damn, it might be exactly the moment to move. But you've gone and let them go. Well, I'll tell you this: You'd damn well better have some idea how to get them back."

Harper tried for several days to entice Tri into reconsidering. But the man was incommunicado and, improbably, immunized against greed. Finally Harper decided to work the flanks. He was angry with Bartlow for pinning the tail on him for a problem of Bartlow's own creation, but he was determined to rescue the operation—whatever the hell it had become—and keep his own record clean. He boarded an Air

America flight to Can Tho and made his way directly from the airport to Tran Van Binh's villa. Later, he wondered whether he had been followed and by whom. There had been signs, perhaps, half-noticed and barely remembered. He rehearsed them again and again during his convalescence, but the details always changed as he tried to remember them.

The sun was setting by the time he got to the villa. The pale-yellow stucco had an eerie glow. The air was calm and close, the way it is before a tornado. The guard at the gate carried a strange-looking carbine, its barrel crudely taped to its stock. He refused to let Harper in at first. But when the boy he sent to the house came back, he brought Binh with him. Binh was full of unconvincing hospitality.

"You are alert for trouble," Harper said.

"Not enough," said Binh, as if Harper's very presence were evidence of a grievous lapse. He showed Harper into the villa and through it to a small office at the rear.

"Something?" he said.

"A beer," said Harper.

He left Harper alone for a moment, then a young woman came in with a bottle of 33 and a glass full of ice. The room was furnished with all the standard items: the gray metal desks and swivel chairs, the file cabinets, a large American map on the wall under plastic. It might have belonged to an FSO or senior adviser. Harper poured the beer over the big, irregular chunks of ice and waited until the carbonation exhausted itself before taking a sip. The girl in the sheer white blouse and black silk pants stood in the doorway watching him. He smiled at her and walked to the other side of the room. There was one shuttered window and a door with a simple latch. He found that the shutters were nailed closed, so he unlocked the door and took a look into the dark. The back yard was ringed around by a high fence and barbed wire. On the ground were perforated steel plates against the mud. The shadow of another armed man passed back and forth at the perimeter.

"Forgive," said a voice. Harper closed the door and fixed the latch. "I have business now. You will wait. I will be soon."

"Sure," said Harper. He was certain Binh would use the time

to check signals with Tri, but there was nothing Harper could do.

"The girl is for you. There will be time."

"There is no need for that," said Harper.

"As you please," said Binh, then he vanished.

Harper gestured the girl to sit down in a chair. She did, shyly. She was not like the bar girls, not yet. Harper watched her nervous smile and tried to get her to relax. Janet always said that if he were ever seduced it would be because he only wanted to be kind. He had always thought she was right, until he met Fran. The young girl tugged nervously at her hands. Harper was just about to laugh when he heard the first shots.

They seemed to come from the street in front, muffled by all the walls. Harper started for the door. The girl began to follow him. He stopped her and made her sit back down. But she looked so frightened that he relented, taking her by the hand and leading her out.

The house seemed deserted. Binh was nowhere to be found. The front entryway lacked the usual guard. Harper opened the heavy door and sneaked a look toward the gate. The Cambodian with the carbine was not there at his post. It was dark, but Harper could see men with rifles congregated near the wire. He closed the door quickly.

"Come with me," he said to the girl. Her hand was damp with fear.

They hurried to the back room where he checked the door to make sure it was still latched. Then he shut and locked the door to the rest of the house.

"You get under there," he said, pointing to the knee space under one of the desks. "I will be over here under this one."

She did what he told her. There were more shots. He huddled down behind the thin, unprotective steel of the desk and listened to his own heart beating. The next shots came from the rear. Then there were Vietnamese voices. Something banged against the back door. He looked up and saw the wood shuddering under the blows. He hunkered down again with nowhere else to go. Then he felt something move behind him and spun around with his hands up. It was only the girl. She crowded in next to him, and he could hear that she was weeping.

"What are they saying?" he asked her. The voices went on, shouting. The rifle butt crashed against the door. She said some words he did not understand and held his arm as if it were life itself.

The lights went out suddenly and then an automatic weapon opened up horribly close. The attackers were in the villa. Harper did not know how long the two of them stayed there on their knees, waiting to be killed, before the voices went away and the silence became impossible to bear. It was the wrong position to die in. He ventured out from his cover in the darkness, and the girl stayed where she was. His hands searched the way before him, and his breath was short and shallow.

The villa was unfamiliar to him, and he moved slowly back to the entryway. The smell of gunfire hung in the humid air. Just as he reached what he thought would be the door, Harper's foot struck something soft, inert. He knelt down and felt for the obstacle. It was warm and damp. It was a man. He pulled out his matches and lighted one. In its yellow flare he saw the torn flesh, the blood. Binh's torso lay tortured by bullets. His head, severed at a stroke, gagged in a final, spent reflex of terror. The wild dead eyes locked on the horror.

Harper led the girl away out the rear door where she would not have to suffer the sight and took her to the home of an Agency man who would make sure she was taken care of. The police were notified, as was the American authority. But Harper did not wait to discuss it. He arranged an immediate flight back to Saigon. The pilots did not like to fly at night, but this was an emergency. He had to warn Tri, to offer him asylum. He did not know which side had done this thing, but he feared it was more than a warning.

What Harper found at Tri's office in Cholon when he arrived there late in the night haunted him ever after. It was not only the carnage—for that was war—but such depraved revenge that it seemed the very image of the rage of the deceived. Tri and his three bodyguards were not decapitated as Binh had been. But each was placed in an awkward self-embrace, their castrated members jammed bloody in their mouths, as if to silence their lies beyond death.

They had to give Harper a sedative that night. They said he had blamed the police, blamed the Agency's lack of care. He had grabbed the elderly night watchman and shaken him brutally, demanding to know what men had come to do such a thing. The government? The Front? The Americans? The Mocking God leered through its primitive mask and laughed.

In the weeks that followed, Harper showed all the adverse symptoms of conjecture. He demanded a full Agency investigation into who had done the killings. This, of course, was out of the question. Nothing was to be gained by such knowledge, and there was a downside risk. If it proved to be the National Liberation Front, this would be useless information, even as propaganda. If the South Vietnamese were behind it, we would be forced to protect them, and the very inquiry itself would strain an already difficult relationship. Finally, we did not need to revive the assassination team stories if, by some strange twist, it turned out that there was an American hand in the murders. This was all carefully explained to Harper, but the reasoning only enraged him. He became, for all practical purposes, dysfunctional.

Not that he behaved in ways that an outsider would consider irrational. He had had quite a shock, and his fear of staying in his flat alone might have seemed altogether appropriate. That there was a lovely woman ready to take him in made it all the more natural. He appeared dutifully at work every day and showed no more than the expected physical signs of distress. But he was consumed by what he took to be his responsibility to resolve the question of the murders. He drafted long, tortured memoranda, laying out the scanty evidence and the surfeit of motives. He made the case against each suspect, one as plausible as the other, and argued that the situation cried out for a conclusion. He was by turns rude, defensive, and sullen. He simply would not accept the solace of indifference that we pressed upon him.

Bartlow, of course, was summarily replaced. His poor judgment had been unambiguously demonstrated. But Harper was indulged for a time. His cables were turned over to the staff psychologists at Langley. The Saigon station was told to carry him, despite his uselessness, like a bad debt. Blood guilt, the psychologists said, had to be allowed to work itself out. While Harper hammered away futilely against his experience, the Agency simply turned away as from a disfigured face. This was the modern view of therapy: the antidote to shell shock was the shell. Give the individual time in the corrupting environment to make his accommodation with it or else he will carry the corruption within him forever. We had to expect our people to suffer some wounds in their work, the psychologists said, and it was our duty to try to rehabilitate them.

This was all very kind and enlightened. But I objected. I wanted Harper cut. Let him go, and if conscience were a problem, give him a pension and find him a new job. But by all means get him out of the Agency. This was my steadfast position.

I had no special animus against Harper. I had never even met the man. But I operated on the principle that the Agency should eliminate all known risks. At the time, he was just a minor functionary. But my view was categorical. His behavior after the murders showed, I argued, a dangerous weakness. It was not simply the fact that he was emotionally shaken. It was the precise nature of his case that was to me so convincing.

He had proven that he was incapable of functioning in darkness. He could not accept the fact that each explanation of the killings was as reasonable as the next, that there was no way to decide among them, that they hung there in the air like a fog. He was, I argued, debilitated by the very environment in which we are condemned to operate. I had no way of knowing, of course, that Harper would eventually spring back and rise as high as he has in the Agency's ranks.

I was overruled. Every man, my colleagues said, has to be granted his one dark night of the soul and allowed to grope his way back into the light. They believed I was being too harsh, too mechanical in my application of the strategy of risk avoidance. At times I think some of them are uncomfortable with the fact that I myself not only can function in the darkness but actually thrive on it like an animal that inhabits caves.

So Harper was given time to recover in Saigon. And by all appearances he did. My objections were forgotten. And Harper was rewarded with a commendation and reassigned to Tokyo for a further period of convalescence.

That night, faced with what he had to persuade Birch to do, it had all come back to him in a rush: the violence, the indeterminacy, the collapse. Harper shuddered in his chair. The low ringing of the wind chimes in his ears, the soft whisper of Tokyo did not calm him. Not at all. But he did not yield this time. He was an outside man, an initiate. He knew the necessity of betrayal, and the cost.

It was a stupid plan they had for Birch, someone's adolescent fantasy. They must be desperate, he thought. Compromise him sexually and the other side will trust him. It was like some barbaric rite of passage. The Slav had seen Birch's love for his wife. So this must be turned to a winning weakness.

Harper wished that he had somehow kept Donna Birch entirely out of his cables. He had thought himself so clever to have used her to show the Slav how dependent Birch had become on the illicit money. He had reported this to Langley. The warped idea had, in effect, begun with him.

There was no question but that he would go along, of course. This was not the same as the last disaster. He would go along because the only thing Birch risked was fidelity. There were no guns. No confusion of enemies and allies. And he knew exactly the means by which he would persuade Birch. Appeals to loyalty, to need, would not be enough. There would have to be an element of leverage. Thanks to Birch's commanding officer, he had the instrument of coercion ready to hand. Birch was terrified of going to Vietnam.

Colonel Robertson had mentioned this when Harper told him of the need to get Birch off levy.

"I'm not saying he's a coward," Robertson had said, "but his wife has a thing about it. He told me he worried how she would cope with the stress of it if he had to go. The Army doesn't exempt men on account of nervous wives. He was definitely in a crack. I felt sorry for him, but there was nothing I could do. You were certainly a godsend for him, Mr. Harper."

Harper sat by the window, practicing the exact manner in which he would use Birch's concern for his wife to overcome his faithfulness and wondering whether all the loftier words men give to their motives—love and loyalty and honor—were not all synonyms for survival. Then he went to the bedroom and lay next to Janet, holding on to her warm, living flesh the way the shy girl in the sheer, white blouse had once held on to him.

11

Birch was not reticent about his sexual inexperience. In fact, he was quite proud of his continence and eager to point out that even before he left Cleanthe this was not for lack of opportunity. There were always the one or two girls in every class at Welton High who had a reputation. How often the infamy was deserved was another matter. The girls who were known this way tended to be those who bloomed early. The sight of that first faint white line of a brassiere under the blouse was enough to set boys to talking. It did not pay to mature out of step with everyone else, because the imagination is a wonderful thing. Ready and able strongly implied willing. Boys wanted desperately to believe. Strangely enough, very rarely did the girls who were known to "go down in the dust for pennies" ever get into real trouble. That was reserved for the nice, mousey ones who sang in the church choir.

In addition to the fickle chance of scoring with schoolmates, there was the stone certainty of the Why Not roadhouse on Route 11 by the creek. There a young man could be initiated in the skinny arms and between the worn thighs of a woman who definitely knew what was up and where to put it when it was.

The roadhouse did not cotton to unaccompanied adoles-
cents; even hell had its rules of admission. So if a boy was
going to experience sin there, he had to be taken by an adult.
To Birch's surprise, some of his classmates went to the Why
Not with their fathers. And to his horror, many of the other
young men envied this depravity. As far as young Birch was
concerned, his father had always been a one-woman man, even
during their separation by war. Not that they ever talked about
such things. It was just assumed. For his father, the subject deserved far less conversation than it got. If a boy on the farm
needed sex education, he could get it from watching the sows.
And if the animals' performance seemed crude and perfunctory, well, there was a lesson in that too.

Birch imagined the inside of the Why Not as smoky and
dark and red with the sound of saxophones. He heard all the
stories in the locker room and behind the barns, some as clini-
cal as the well-worn anatomy text in the town library, others as
suggestive as the paperbacks you could only buy at the grill
out on the highway where the Trailways buses stopped. He was
curious enough about it to cruise past the Why Not on idle
evenings when he had the use of the car. Dingy and set back
from the road, during the day it seemed about as erotic as a
silo. But by night it glowed, and if you looked closely you
could see shadows against the upstairs blinds and imagine, just
imagine.

Occasionally the Why Not girls braved the clucking propriety of the Welton folks and ventured into town to spend some
of their ill-gotten gains. Their sin gave them an air of superiority and disdain, marching down the main drag in pairs, heads
turning neither this way nor that since there was so little in
such a backward place that women of the world would possi-
bly want to see. In the barbershop the men went draped in
white and foamy-jowled to the big window to watch them pass.
A daring customer might even mention a name. The police
chief eased the force's single cruiser to the curb and waited
there until the ladies had done their business, just to be careful. For the stern matrons toiling away in the church base-

ments, making doodads for charity, there was no need to strain for conversation. They were outraged, enthusiastically.

One day Birch and a gang of buddies were hanging out in the five and dime after basketball practice when the Why Not girls flounced in.

"Oh Jesus," said Jimmy Tucker.

Even in the lush glow of their notoriety, they did not seem to Birch attractive, or even mysterious. One had acne scars. Another was plump in the wrong places. A third spoke with the voice of a ripsaw. They all seemed too pale, as if all the old warnings about how the bad thing sapped a person's strength had some truth. Of course, they weren't exactly dressed to kill in their car coats and sweatshirts, their hair up in curlers under cheap scarves. But still, their skin was shiny and scaled, their lips as red as a sore.

"That's her all right," said Jimmy Tucker. "I seen her tits."

The other boys challenged this incredible boast, but Jimmy Tucker remained adamant. They shoved and jostled their way to the front of the store in an elaborate effort to get a better view without being conspicuous. Someone said the girls all seemed pretty old. Someone else explained that you had to be of age. It was some kind of law, he thought.

"Which is the one, Jimmy?"

"The one in the green over there by the candy. Oh Jesus, she sees me."

She was tall and skinny with a long chin and nose. But she wasn't exactly ugly. She might even have been pretty, if she cared. Birch watched the girl and tried to imagine her wrestling with gawky Jimmy Tucker, his mess spurting up inside her. It was simply implausible.

"What were they like?" asked one of the others. "Her tits I mean."

"They moved around," said Jimmy Tucker. "It was kinda dark."

"Go talk to her. Maybe she'd like a vanilla Coke."

"I got some cigarettes hidden at home. Maybe I should go get them."

"I think they can buy their own," said Birch.

Meantime, Jimmy Tucker was getting a little nervous.

"Maybe I got her in trouble," he said.

"She don't look it."

"You don't want to let them know who you are, just in case," said Jimmy Tucker.

"What's she gonna do? Tell your old man? He was probably with the fat one in the blue."

"Hey. Cut it out. I told you not to say anything about that," said Jimmy Tucker.

He tried to squeeze his way to the rear of the gang, but a dozen hands thrust him front and center. Then the girls finished their business and moved toward the door the boys were guarding. Jimmy Tucker braced himself and moved forward to greet them.

"Nice day," he said boldly, with only a slight squeak.

But the girls all brushed past him, smelling of flowers and smoke, like a wake, and left without even so much as a wink.

"Are you sure that was the one, Jimmy?"

"How could I forget?"

"She did," said Birch.

At least that was how he remembered the incident to Harper as he pleaded why he could not go along with Langley's plan. He was against this sort of thing, always had been. Birch did not mean to seem prudish. It wasn't that he condemned others. He never said a thing to the young fellows in his unit who liked to have a fling once in a while. That was up to them. But the idea always turned Birch off.

"We're not asking you to enjoy it," said Harper.

"You don't understand," said Birch. "I have a wife to worry about."

"You aren't going to be taking her along with you."

It was the damnedest thing the way Harper turned everything around on him. A mission of "extremely high priority" had come up, he said, and the Agency was considering Birch to carry it out. Duty always means sacrifice. Without men who are willing to go the extra distance, a country is in trouble.

"Sure you find it offensive, Birch. I would have thought less of you if you didn't. But when you refuse—when you scruple the Slav in this—it makes him nervous about you."

Harper made it into a question of loyalty. And while Birch thought of it that way too, it had always been in the opposite direction: honesty with his own strongest feelings and fidelity to his wife. Harper blurred the borders, made things into their reverse. Birch did not need to be persuaded that the cause was just and significant. He was never one of those who ran down the country or took the Russians lightly. But this was all so dirty. Why should his fidelity matter to anyone but Donna and himself?

"Faithfulness is an attitude, not an act," said Harper. Birch told him he had no idea what that meant.

"It means that duty all comes together. When you do what you must, you are keeping a trust, not violating it. If she could know all the facts, she would understand that what you have to do does not in the slightest way change the integrity of your heart."

Birch resisted. But he had no arguments to turn Harper back; he was not a philosopher who could twist words inside out. He simply could not imagine that anyone really took this idea seriously. He had faced the question many times before. There was always someone trying to loosen him up a little, to show him a good time. Maybe he had seemed a little uptight in refusing, but it didn't bother Birch anymore. It was just the way he was.

"No way, Mr. Harper."

But then Harper took the extra step. It had been decided, he said, that either they went through with the plan as described or the operation would have to be terminated. If it made no progress, it was just a waste of time.

"Your personnel file has been on administrative hold," said Harper. "You have been frozen in your current assignment indefinitely. And if the operation were to progress the way we want it to, we could not risk a later compromise by allowing you to go where there might be a threat of capture. But if we terminate, your files will be freed up."

"You mean Vietnam," said Birch.

"Wherever. The Agency would no longer have a legitimate interest."

"That's not fair."

"I'm just stating the facts, Birch. It's your call to make. I want you to make it fully informed."

Birch was desperate for another way. They could offer the Slav more than he was asking for. They could demand more money. Birch could pretend to be drinking too much. He could talk about drugs or say he was queer. But Harper did not seem to be listening.

Birch left the safe house without agreeing to anything. He was confused and angry. The Tokyo streets had never seemed so alien to him. Swarming crowds swept him along. He had no destination. Glaring neon signs made rude demands upon him, and he could not even understand the alphabet. He found himself on the night streets that had always repulsed him. Touts grabbed at his sleeve. Pimps whispered their wares. He shook them off, but he could not bring himself to leave. He was not drawn to this vile place; he was thrust into it. Duty, compulsion, faithfulness, it was all in the angle of vision, the points obscured. But no matter how hard he tried, he could not get any one of them to resolve.

Donna was waiting up for him. As he fumbled his keys in a trembling hand, he heard her footsteps on the stairs. She opened the door. He never came upon her by surprise. She had ears for his returning. When she kissed him, he received it as shame.

"Before I forget," she said, "Mr. Nowicki has been trying to get in touch with you. I didn't think you had duty, so I didn't know where to tell him. He said he wasn't where he could leave a number."

"That's all right," said Birch. "He never is."

"He seemed kind of urgent."

"I hope he didn't bother you too much."

"I wasn't doing anything special," she said. "A letter came from my mom. I was trying to answer it. Dad caught the flu, but he's getting over it. They had to practically strap him to the bed. You know how he is. I thought we could send them one of those cute little wooden dolls they sell. Don't you think that would be nice?"

"It would be fine," he said. He could not meet her eyes. She did not even ask what had kept him so late. And this night her trust was like a reproach.

"Mr. Nowicki asked whether it would be all right to call again," she said.

"It's late."

"I told him it didn't matter when. Have you had your dinner?"

The cat peeked around the corner at him and Birch kneeled down to let it run its arched back up against his hand.

"I'm not really hungry," he said.

"I've already nibbled," she said, "but I could make you something quickly."

Birch ran his fingers gently down the kitten's neck to where the little bell hung to save the birds.

"I think you're making progress with her," she said. "She used to be so frightened. Why don't you go upstairs and get into your robe. I'll bring you something nice. Go ahead."

Birch managed a smile and gave her a peck on the cheek, a small enough betrayal. He remembered Harper's argument: "If you continue the operation there is no reason for her to know it and every reason for her not to. But if we terminate and you have to go away, there is no way you can save her from that." He climbed the stairs to their bedroom. She was terrified for his safety. One of her cousins had been killed there. She was sure Birch would be hurt. Birch, of course, tried to kid it. He was too valuable to be sacrificed to the guns, he said. And he had even begun to believe this himself, once he was told it was true.

The telephone crouched on an end table next to the bed. Birch picked up the receiver and held it in his lap. When the dial tone turned to a warning siren, he choked off the noise under a pillow until it was dead, then laid the instrument down next to its cradle. For a moment he felt relieved. He had done something to protect them.

"Are you OK, honey?" she asked as she set a tray of fruit and salad and cheese before him on the bed.

"A little tired is all."

"Don't you want to tell me?"

"I thought I just did."

She did not react to his sharpness. Sometimes she could be so dense to him, shutting out even so much as he could share of the things he had to hide.

"It's that new lieutenant, isn't it? He's been getting on your nerves."

"I don't care about him," he said.

"You'll get him broken in. You always do."

"I haven't even spoken to him."

"Jerry, did you know you left the phone off the hook?"

"Leave it. Just leave it there."

"I think it runs up the bill, honey," she said, reaching for the receiver. He grabbed it away from her.

"Just do what I say for once."

"You said you wanted to economize," she said. "That's the only reason."

"I know what I said."

"Mr. Nowicki may be trying to call."

"Maybe I don't want to talk to him. Maybe I don't want to talk at all. Have you thought of that?"

Donna crossed the room and sat in a wooden chair, sat there like a wallflower, too proud to show that she had been hurt.

"Say what you're thinking," he said. "Go ahead and say it."

"I don't know what to do when you get upset, Jerry."

"Maybe that's the problem."

"Tell me what I should do. Can I rub your back? That sometimes helps. You take things so seriously. You get so tense."

"My back is fine. It isn't my muscles."

"Oh, Jerry."

"You just sit there and take it, don't you? You just let me do this, and you don't even know why."

"Please, Jerry. Maybe if you eat something."

"Oh, that will take care of it, all right. Fill my belly. Rub my back. That ends it, doesn't it, your responsibility."

"You shouldn't go to bed hungry."

"I'm not your baby, Donna. There isn't any baby. We can't have one. You'd better get used to the idea."

She stood and walked past him to the door.

"Where are you going now?"

"I'm going to get ready for bed," she said. "You do whatever it is you think you want."

"I'm sorry," he said, but she was gone.

He picked up the tray and started downstairs with it. He stopped outside the bathroom door, leaning against it and listening to the water in the sink.

"It's not you, Donna," he said.

"I never know," she said.

"Well, it's not."

In the living room he tried a piece of cheese. It turned to dust in his mouth. The beer was bitter and warm. He sat in the dark hearing Donna's slow, sad movements above him. He had wanted her to know what he sacrificed for her, and the secret corrupted him. It forced him to trouble her most tender spots. These he knew too well, the emptiness of their failure to have a child, her desperate need to make him happy. There would always be ways to make her share his problem. And they would always be cruel. He hated the weakness in him that could not bear this thing alone. It had all become so turned around. Love was insult, infidelity a kind of duty.

Her sounds had stopped, and he found himself weeping in the silence. If he did not do what they wanted of him, it would only get worse. He would get his orders. They would have to part. A year away from him. And it would not be one of those things that just happened or did not happen like getting pregnant. She would want to know why. And he would know, know the fact that he had brought it on himself to save his miserable conscience. But he would never be able to say.

He went to her. She was breathing uneasily, not asleep. She had had tears too. The telephone remained off the hook. She had left it that way, not caring about his reason. He picked it up and replaced it gently. Then he leaned over and held her.

She did not move away. He held her there, sheltering her with his body against all harm.

"I love you, Jerry," she said. "I don't know what I've done wrong. But I love you."

"I'll make it better," he whispered. "Don't worry. I'll work it out. It's my fault."

Back downstairs in the kitchen, he turned on the radio to the Armed Forces Network to cover his voice. Then he drank another mouthful of beer and dialed the number Harper had given him. When the man on the other end answered sleepily, Birch said he was trying to call for a taxi. "Good," said the voice. "I knew we could count on you." Birch hung up and sat down to wait for the Slav to call.

Once the date had been set, Birch savored his sense of guilt. Donna did not realize what he was apologizing for as he treated her so penitently. She thought it was for what he had said about a baby. But in fact, he was making amends in advance, and the very distastefulness of what he had agreed to do was the only thing that made it possible for him to contemplate doing it. It was a sacrifice, he told himself. And, in a sense, it was like the counterfeit treason of turning over secret documents to the Slav. The papers were authentic, all right. To an outside observer, he would appear a disloyal wretch. And every time he did it, Birch felt a tightness grip his belly at the plain treachery of it. Yet still he knew that the treachery was all the other way. The hidden motive absolved him, not because it was beyond his control but rather because he had freely and blamelessly chosen it. He was responsible for his innocence, and he suffered for it.

The suffering was important. He knew he would get no pleasure from the act. He would be disgusted by it and by his own choice of it. It would be just as false as the lies he told the Slav. If Donna knew, she would be horrified. But she would not know. And because of that, she would in fact know only the truth. She would have him as he really was, and all he would keep from her was the deceit, the false and fault-ridden Birch who never was.

At the restaurant the Slav had another assignment for him. Birch had trouble keeping his mind on it, but the Slav did not seem to notice. Birch had notified the Agency of the meeting place by the usual method. The dead drop always worked, and Birch could feel allied eyes upon him. What the Slav wanted this time was more of the same. Birch was disappointed. They still did not take him seriously, and he had no excuse to balk. The material would be simple enough to provide. It was as if they were still testing him. But testing him for what? Birch was risking everything, and yet he sometimes felt that the Slav only cared to have a reason for taking fancy meals.

Birch translated his frustration into a convincing reluctance. He worried aloud to the Slav about getting in too deeply. But the Slav reassured him that it was all just bureaucratic silliness. Why shouldn't a firm selling to the United States Government have the information that would make the products first-rate?

"I don't know the reasons," said Birch. "There must be reasons."

"But you will try, yes?" said the Slav.

"I'll see. It makes me nervous."

"This is not good. The company does not like its high-class friends to be upset. It wants them relaxed and happy. You will come with me tonight and take the mind off small things."

"I'm game," said Birch.

"Good. Excellent."

Birch thought again of the pale girls of the Why Not roadhouse. His face was pinched with guilt. And he was grateful with shame.

They took a taxi to the place the Slav had selected, and from the outside it did not disappoint Birch's expectations. They entered, the Slav wavering under the assault of all the whiskey and wine, and Birch paused to look at the doorway framed by faded, tawdry photographs of the ladies. The doorway led to a dank staircase, and the Slav snaked his arm across Birch's shoulders as they climbed. At the top, the door was painted an outrageous pink. The Slav knocked and an old woman opened it.

"Ah, Mr. Nicki-san," she said. "We wait for you always."

The Slav bent at the waist in a clumsy parody of a bow. She took him by the hand and led him inside. Birch followed.

It was not at all the way he wanted it to be. The room could have been someone's parlor. It was bright, much too bright, the furnishings sharp and traditional. The old woman directed them to a table with a nice linen tablecloth. In the center a candle burned unnecessarily. It gave off the faint aroma of sandalwood.

"Your friend?" said the woman.

"Friendship needs whiskey," said the Slav, "needs a woman."

"So," she said and padded away from them in the old-fashioned silk kimona and obi.

Where was the leering darkness, the smoke, the sullen sound of saxophones? Birch could not connect up this place with his sense of sin. He should have insisted on choosing the place himself. He should have taken the Slav to an R and R bar, the one beneath the safe house, where there would be no confusion.

The Slav sat heavily, fist braced against the table, head sagging down. He was humming a hoarse, tuneless song, full of passion.

"Friendship," he said when he noticed Birch watching him, then he resumed his lonely melody.

Birch heard giggles from the back where the old woman had gone. It was the young, girlish laughter of complicity in the knowledge of the foolishness of men. Something stirred in him at the sound. He put it down to fear, a boy's fear of being discovered at his pleasure behind the door. It was the same feeling that in his youth had both drawn him to the forbidden thing and held him back. It was the two-edged rush of sensation that made him want to hide his desire and that at the same time gave desire its special savor. It belonged in darkened rooms. It was out of place in company, an adolescent's erection springing up beneath his suitpants in church. Birch shifted restlessly in his chair.

"You will want Toshi, Mr. Nicki-san?" said the woman, shuffling back into the room.

"She moves like the water," moaned the Slav. He was off somewhere in a place reachable only via alcohol and loss.

"Toshi!" said the old woman, clapping her hands together sharply.

The girl emerged with a bottle and two glasses on a tray. She was small and smooth as a child, but she had the full curves of a woman. She was dressed in white, Western-style, lace. Her bright black hair hung gently to her shoulders, straight like Donna's, shining. Birch stood to greet her and wished she were less than what she was.

"Ah, Toshi," said the old woman. "See who is here."

The girl smiled and averted her eyes from Birch. Then she set the tray before them, leaning down so close to him that he could smell the light, exotic spice. Her small, firm breasts rippled free beneath the cloth as she moved. The Slav put out his meaty hand, and she moved against it, obliging and shy. Birch had to look away, for what welled up in him was more than pity.

"Mr. Nicki-san," he heard her say.

"Is my friend here," said the Slav. "Jerry Birch."

"Jell-y," she said, struggling sweetly with the sound.

"Here is another," said the old woman as a second girl entered the room. "You like, yes?"

She was taller, more ample. Perhaps she was just as young, but this was not something measured by years. To Birch she seemed, mercifully, as hard as the past. Her gray blouse was translucent in the glaring light, and the nipples stood out like tarnished copper pennies in the dust.

"She is Kokura," said the old woman. The girl sat down next to Birch and placed her hand lightly on his thigh.

"That's fine," he said. And it was, because the pressure of her hand spread numbness where the sight of Toshi had made him come alive.

"I leave you," said the old woman, and she padded out of the room.

Toshi lifted the bottle and carefully poured the liquor into the glasses.

"Do you speak English?" Birch asked, and Toshi looked up at him and giggled.

"She country gul," said Kokura, her voice deep and dark. "I know words."

Her hand moved on him as she slid closer so her full breast rested softly against his arm. Toshi finished pouring and put the bottle down with a small flip of the hand that charmed him.

"Moves like the water," said the Slav, leaning over and taking her head between his hands for an awkward kiss. When he had finished, she shook the tangles from her hair and smiled as if she had never suffered a touch. Then she reached across the table and pushed Birch's glass closer to him.

"Doesn't know a damned word," said the Slav. "But beautiful sounds she can make."

Birch sipped the biting liquid and closed his eyes against its fire. He was confused by the powerful urge he felt to protect her, the more by other emotions he wanted to deny. He took a deep breath and opened his eyes to the light, which warmed her face.

Kokura laid her head against his shoulder and whispered, "You happy in me?" Birch was sickened by the thought. Across from him the smaller girl accepted the Slav's murmuring advances in silence. His hand cupped her breast. Birch searched her face for pain. But there was nothing. She saw him watching her and smiled as if there were no evil in the world.

"You no like Kokura?" his partner whispered. He took another drink and then touched her hand where it rested high on his thigh. She took it and placed it flat against the warm skin beneath the hem of her skirt.

"Soft," she said.

"Yes," he said, but the touch was nothing.

Suddenly he wanted it over with. He wanted to be away from Toshi, to finish this thing with an urgency, not of flesh, but of conscience. He wanted to be alone with the one who had no power to move him. His body would respond, but that was not the truth. That was no more than the most meaningless animal reflex. He could not allow himself the feelings Toshi was igniting in him because that was the truth, and the truth was betrayal.

"Is good place," said the Slav, gulping down his drink. "Is for men, for friends."

"I think I'm ready," said Birch. Kokura understood and slid harder and higher against his fingers. He left them where they were by the sheerest act of will. "Are there rooms?"

"Here," said the Slav, lurching forward across the table. "Toshi is for you tonight. She moves like water. I take the other. Everything share."

"No," said Birch. "No, please."

"Is what I want," said the Slav, raising his glass in a sad, soulful toast. "For friendship."

"Kokura is better for me," said Birch as she slipped away from him. "I like her. I really do."

But the Slav was having none of it. Already the tall one was on his lap and Toshi was gently guiding Birch to the door. Even the light touch of her hand in his thrilled him. He tried to deny it, but it was irrefutable.

The little room she led him to was just as bright as the parlor. It was small, but pretty, decorated with arrangements of dried flowers and gentle nature prints. She lighted the incense then sat him on the large bed. He wanted to stop her, but he neither knew the words that she would understand nor the reasons that would give him strength. Before he had a chance to say anything, she had pulled her dress off over her head in a single, dancer's movement, and she stood before him, young and soft and strong as yearning. He turned away. The light. He wanted darkness. But she was luminous. The small black triangle low on her belly was the only shadow, and it was a glorious mystery. He shuddered at the wonder of her nakedness. Then she moved closer and took his head between her gentle hands and carried it slowly to her breasts. He yielded to them, yielded even the self-deception that had sustained him. Then she eased him down upon his back and moved to caress. He allowed it, wanted it, her skilled hands touching him there, and then her lips sheathing him where no other woman had ever gone.

When he broke away, it was because he wanted her.

"I can't," he said. "I just can't."

From the bed, she looked up at him confused. He wanted to comfort her, to explain the convoluted truth she had touched off in him. But it was hopeless to use words. He bolted from the room.

The Slav was still preparing himself outside. Kokura looked up, startled, her hand deep in his pants, as Birch darted past. The Slav tried to stand, but he was hobbled. He fell back in the chair.

"Wait, friend," he bellowed. But Birch was already halfway down the stairs.

12

Intelligence is string-saving. It may be intriguing to imagine us haunting sullen streets or slipping over armed borders—and there is some of that, to be sure. But chiefly we are collectors of paper. We never throw anything away. Take the case of coded texts. When we have no idea how to read the gibberish, we still keep everything we intercept against the day when we break the cipher. It becomes almost pathological. During the purges it came out that the Sisters even kept documents so sensitive that they were supposed to be destroyed. They called it the Do Not File File.

In cases such as CONVERGENCE the documentation is particularly rich, and it provides not only a wealth of transcripts of surveillances and interviews but also a running commentary in Harper's memos to file. Of course, it is sometimes said that every man is the hero of his memos to file. There, in the security of a manila envelope nestled in the heavy steel drawer of a safe, a man can square the circle of principle and practicality, say the things he wished he had said, write for history—in several alternative drafts. But Harper's memos to file from this period are remarkably consistent with the cables he sent to Langley. He warned that the plan to shake Birch loose of the Slav would in the end destroy Birch's usefulness to us. Even if the maneuver went off as planned, he argued, it would be impossible to reestablish Birch's total confidence.

Meanwhile at Langley, opinion shifted as soon as word reached us that Birch had balked. The unanimity that had originally supported the plan evaporated. Those who had not committed themselves in writing suddenly swung to Harper's view and maintained they had shared it from the start.

I do not respect a man whose mind is cut to fashion. I date back to an earlier era when careers depended more upon the strength of a man's word and will than upon his agility in leaping on and off the moving wagon. To me, it is a mistake to rush to a reassessment under the impulse of each new fact. Reality is far more constant than it serves our momentary interests to have it. So, too, is character which, when flawed, is broken at the hard places, down to the very marrow.

In short, I agreed with the plan and respected Harper's disagreement with it. Some of my colleagues wanted to take drastic action after the brothel episode to salvage Birch's connection with the Soviets. He was the best we had at the time, and it was feared that he had broken it off irreparably. I took the position that it was best to stand pat and see what developed.

We do not know precisely what transpired within the Soviet hierarchy in the wake of that ludicrous evening. In general I profoundly disagree with Harper's dream that all adversaries come together as they rise. But I would be willing to grant him this much: that bureaucracies have their own imperatives that transcend ideology. So I assume that the Soviets were doing just as much backing and filling as we were at Langley. However seriously they may have taken Birch's potential, they would not have wanted simply to discard him. If intelligence agencies are loath to part with scraps of paper, how much more so with living flesh.

Birch dutifully cleared the dead drop the next day and learned where his next meeting with Harper was to be. He showed up punctually and went on at some length to explain the convoluted emotions that had animated him. He said he was sorry if he had failed in the mission, but he was relieved to have made the choice he had. Harper had been right about the importance Birch gave to the saving illusion of continence. As Harper chatted with him—and that was really all it was, since no longer was there anything immediately to be done—he had the feeling, duly reported, that the incident had left Birch terribly confused. Birch wanted to tell the truth about what he had done, but he had trouble separating recognition from self-deception. He remembered what had happened, what had been said, with the usual remarkable accuracy. But when it came to expressing the sentiments that led him to break away and risk the very thing that he had agreed to compromise his principles to avoid, he could not put it plainly. It was as if, Harper reported, Birch had been so shaken that he had lost touch with himself. He was only clear about the lie.

No one paid much attention to Harper's report at the time, though later Harper himself began to place heavy emphasis on the sketchy psychological observations he had recorded there. All that mattered initially was that Birch seemed more than willing to go on with the operation if the opportunity presented itself. His attention, like ours, focused on the Russians.

And they were silent. The Slav made no attempt to renew contact. He was nowhere to be seen. Our listening post at the brothel picked up nothing but laughing Japanese patrons out on their expense accounts with business clients. The stakeout at his residence came up blank. We figured that either he had himself gone into hiding (which seemed unlikely, though we discreetly put out the word that should he offer himself up as a defector, he was to be politely but firmly turned away) or that they had hustled him off to a safe house while the damage done in the debacle could be sorted out. It was not, however, until the Slav showed up at the airport in the company of his GRU superior and several burly KGB musclemen in their mirrored, wraparound sunglasses that we allowed ourselves any optimism. We did not have the resources to tell us whether someone was going to pick up where the Slav had fumbled, but it was clear that at least one element of our plan had been a success. The Slav, pushed onto an Aeroflot jet and looking none too happy about it, was out of the picture.

Since that time we have seen nothing of the man who called himself Nowicki. There is no evidence that he suffered the same violent end as his father. The Soviets have lost their taste for such spectacles. Among some elements of our national security apparatus, this is credited to our encouragement of the Kremlin's better side. They seem to think that the Soviet Union is a troubled adolescent who only needs our firm but understanding guidance to help it take its place as a mature and functioning adult in world society. The psychoanalytic theory of foreign policy is all too fashionable now. The bear lies back on the couch and acts out its surly fantasies while we stroke our chins and give out knowing grunts. We create dependency, watch for abreaction, reward behavior we find suitable, and hope for the best. Fortunately, I am by no means alone in believing that the patient is irredeemably incorrigible. If the Kremlin no longer hides its failures in elaborate false confessions and no longer executes the functionaries who have fallen from grace, this is not because the Soviets have developed a distaste for barbarity. They have simply become cleverer, absolutely sardonic in fact. Nowhere more so than in their switch from concentration camps to mental hospitals. Now they commit their problems to the schizophrenic wards rather than to the firing squads. But is this anything more than a crude joke on us? If the West treats the Soviet Union as a problem in behavior modification, then the Kremlin will treat its own recalcitrants to the same humiliation. When they parody our weaknesses, they also serve their own iron will. This is convergence as ridicule, as scorn.

If I had to guess where the man who called himself Nowicki ended up, I would guess a large, impersonal institution. His diagnosis: sexual indiscipline dysfunction. The etiology: lack of father figure during boyhood and the corruption of the West. Treatment: isolation, medication, self-criticism. He is probably encouraged to keep a journal, which is routinely read and analyzed. By now, I would imagine, he has probably managed to torture his memory sufficiently that, like his father before him, he has reconciled it with the inevitable and infallible judgment of the state. This would be the true nature of his madness.

When the Slav was spotted leaving Japan, Birch was alerted to be ready for a new approach. Harper could not predict how it would come, if it did, nor who would make it. But he warned Birch that he should not show any ambivalence about the incident at the brothel. No explanation of the unexpected attraction he felt for Toshi, no breast-beating about his sense of guilt. Leave the Why Not out of it. He was to tell the man that the Slav trapped him in a situation that disgusted him, that he tried to get through it without a confrontation, but that he found he could not. "Be clear about your resentment," Harper told him. And Birch, for his part, seemed relieved to be given a script.

Our period of uncertainty was mercifully short, and to the surprise of the trimmers at Langley who had rushed in to criticize the plan they had earlier endorsed, the Soviet response was better than anyone had the courage to predict. The Russians made their move boldly, the way they often do when they feel themselves cornered. Without regard to all that earlier business about bidding on U.S. contracts and the European firm, they caught Birch in public in a manner that they must have known would alert him to the real nature of the enterprise.

Birch was in the city shopping for an anniversary present for Donna when a well-dressed man with a donnish accent and a cultivated manner appeared next to him at a jewelry counter and addressed him by name. Birch did not have to strain to seem surprised. The store was aswarm with activity, and the face that addressed him was altogether unfamiliar to him. But not to us. It was Anton Ignatyeff Kerzhentseff, the KGB's most accomplished agent in Asia, the man credited with the most serious recent compromises suffered by the British and French. He was also believed to be the prime mover behind Black Body.

Kerzhentseff introduced himself by his real name and said he was Nowicki's successor. He apologized for the Slav's unforgivable boorishness and said he wanted to reestablish a working relationship. Birch was reluctant, as coached, with a nice grace note of discomfort at having the discussion at all out in the open. Kerzhentseff picked up on this immediately and assured Birch that there were better occasions to discuss the details. He suggested a place to meet later in the day; a car would pick up Birch at a busy street corner in the Ginza, and they would have a real talk. Birch agreed. With that, and nothing more, Kerzhentseff vanished into the crowd.

Birch went uncovered to the meeting place and waited. We wanted to take no risk of revealing our hand. He idled nervously for the better part of an hour, watching the silent images flashing on a dozen television sets in a display window until a Mercedes pulled up to the curb and a deep voice spoke his name.

Birch slid in on Kerzhentseff's left in the back seat. The driver spoke a few words in a foreign tongue; Kerzhentseff replied. The car slipped into traffic and simply wandered. Their language was Russian, and all pretense was off.

"You are with Mr. Nowicki's firm in Europe?" said Birch.

"Surely you have come to be skeptical of that," said Kerzhentseff.

"I don't understand."

"Perhaps it has been a comfort to you that we have until now persisted in the rather transparent story about our business."

"What do you mean?"

"Come now, Mr. Birch. You are an intelligent man. You know what we have been suborning. You recognize the nature of what we have asked you to provide. Have you really been blind about who we are?"

"I suppose not," said Birch.

"There is no need to be alarmed," said Kerzhentseff, lighting an acrid French cigarette and watching the biting smoke swirl toward the vent. He was a small man with delicate features, delicate the way a razor is: the finer the edge, the cleaner the wound. He held his right hand immobile on his knee, the fingers in a stiff and useless grip. "We must move our relationship onto a new plane. A new candor, Mr. Birch. You are quite safe with us. In fact, you ought to feel the more secure now that Mr. Nowicki's stupidity has been eliminated. The greatest danger is an ignoramus."

"Who are you then?"

"Candor, Mr. Birch. I am exactly who I say I am. And since it would be imprudent to tempt you to inquire, I am the representative of my government, the Soviet Union."

"I think you should let me out now," said Birch.

"Please," said Kerzhentseff. "Do not feel obliged to register shock. Your protection will not come from a false innocence of heart. It will come from us, if it is to come at all. And, Mr. Birch, I think you will find that a continued relationship between us will also bring you certain other, more affirmative rewards. Remember that your risk will not increase by further cooperation with us, only by a rash decision on your part to break off. We have evidence of your past crimes in your own hand. You have already taken the largest step. It would be simple enough to destroy you. This very night. Too simple, Mr. Birch. Only a fool like Nowicki would even consider it. We have no such inclination. You can be of considerable service to us. And, as you see, we to you as well."

"I need time to think," said Birch. "It's different now."

"But you do not yet have anything to think about, Mr. Birch. Let me provide you a framework which perhaps you will find reassuring.

"First, you may be sure there will be no further embarrassing episodes of the sort you suffered with Nowicki. I see you have a package. I trust you found something suitable for your wife. Donna, isn't it? In the future you will be able to give her even finer things. Otherwise, there will be no connection whatever between what we do and your personal life. It was a mistake for Nowicki not to respect your feelings in this regard. Terribly unprofessional. We are men of the world, Mr. Birch. We know the need to take people as we find them.

"Despite what you may have experienced, we are simply engaged in a business like any other. Every business has a few unfortunate individuals it has hesitated, out of reasons of kindness, to weed out. In my work there are, of course, certain requirements of care that give it a shadowy and unbusinesslike cast. In fact, you will find that I am far more strict on these matters than Nowicki was. It is for your own safety as well as ours. The telephone will not be used. At times, as I am sure you will understand, we may have to take steps to authenticate what you tell us. You may at first think of this as an ordeal, but I assure you that our techniques, should we have to use them, are sophisticated and painless. If you hide nothing, you will have nothing to be concerned about. Henceforth, we will communicate primarily by dead drop."

"What?" said Birch.

Kerzhentseff watched him closely and then smuggled out a small smile, or what passed for one.

"An odd phrase," he said. "It is merely a way of speaking. It has nothing to do with corpses. I will explain in a moment. The work is really quite bloodless, I assure you, Mr. Birch.

"Finally, I must explain that what we expect of you has nothing to do with loyalty. You are not in the slightest way interested in the great ideological debate between our nations. And frankly, neither am I. It is merely a way each has of asserting the abiding fact that it is what it is and not the other. If the dogma of Marx had not intervened, it would have been some other—a great religious schism, perhaps, or a feud of the blood. We are fated to be world adversaries, Mr. Birch, and how much more harmless that competition is when it is undertaken quietly in the darkness than with the awesome weapons both nations possess. I think of our work here as the last, best alternative to battle.

"I hope you will not think me too much a cynic when I say that for small men such as we are, the only loyalty we dare entertain is to the self. We look to our own advantage and try to remain safe. We do our best for the greater peace—which is a standoff—by pursuing our own individual self-interest. And if that sounds dangerously close to the doctrine of capitalism, so be it. I must say that I have less faith in the so-called invisible hand, however, than I do in the real one of flesh and blood which I hold out to you as a sign that our interests are mutual."

Birch hesitated, then took the strong left hand and shook it with his right in a strange and serious embrace, a communion.

"I am a man of my word, Mr. Birch. I expect the same of others. And it is an expectation that I am willing to enforce. Together, however, we can accomplish a separate peace."

One has to admire Kerzhentseff's instincts. Though he might have approached a European differently, he did not use ideology to seduce Birch. Quite the contrary. He denied it. He is a shrewd man, this Kerzhentseff. And Birch recognized his power immediately. Kerzhentseff is a man who shares my belief, contra Harper, that while some people may be turned by the illusion that the Soviets offer some special difference, most Americans need only to be persuaded that beneath everything we are really all the same.

13

Harper acknowledged Birch's impression that Kerzhentseff was much more formidable than Nowicki had been, but he did not provide any details from our extensive files on the man. Harper did not want to scare Birch. One has to have gone a few rounds, quite a few, before he achieves the respect on the other side of terror. The bear's bulk is daunting, but it can bring him down hard. The claws are sharp, but they can be evaded. Harper merely tried to bring home to Birch the essential assumption that even the cleverest opponent can be fooled.

There was another reason for not sharing all we knew with Birch. As later events demonstrated, an asset must always be treated as a potential double. The fewer weapons you give him, the fewer he will have to shoot you with later. We are as jealous in guarding the information we have gathered about our adversaries as we are of our own sources and methods. In many instances the one can lead to the other. Every fact leaves a trail: individuals who have been entrusted with it, channels through which it has been communicated. Whenever our opponents learn that we have discovered something they have held closely, they make a map of its history. And when trails begin to intersect, somebody can find himself standing out in the open with nowhere to hide.

I have always been curious—perhaps it is vanity—to see the file they keep on me at Lubyanka. Surely they must by now think they have discovered my true identity. But do they know my childhood? What drives me? Have they seen any of the foolish stories I wrote when I was in school? I wonder if they know that I wince now whenever I think of those fables and remember how much of myself shone through the amateur, transparent lies. Do they know how close they came once in Nuremberg to destroying me? Or that I was the one who found their men in the NSA? I wonder whether they have the wit to fear me.

We had ample reason to fear and respect Kerzhentseff. One of his trademarks, since he took his first overseas assignment, has been to work in full view. He travels in polite society and makes no attempt to discourage—beyond the perfunctory, laughing denials at cocktail parties—the rumors of his clandestine affiliation. He is a cultured man, widely read, a catholic intellect. He speaks flawless English, learned during his childhood in Britain where his father was assigned as a midlevel diplomat. If anything, the whispers about his KGB connection only make him more attractive as a guest and confidant of all the sophisticates in the press and public service so bored that they seek out danger as if it were a fashionable new nightspot. This has given him entrée into the whole, bright, nationless world of society and scandal. But what is not fully appreciated by those who share their gossip with him and hope vainly for something fresh and evil in return, is just how far up in the hierarchy Kerzhentseff stands. He is a colonel in the KGB's First Chief Directorate, the organization dedicated to the harassment of the West. And it must be understood that in his service a man could not become a colonel by time in service, dull competence, and the ability to flatter a superior's wife. In the KGB a colonelcy is a measure of brute cunning.

Kerzhentseff has had many opportunities to demonstrate this capacity. Of his earliest career in the Red Army during the closing months of the war we know little except that he received a commission—his father's position probably guaranteed that—and a wound. A bullet tore into his right arm at the elbow, leaving it forever useless. He has compensated by strengthening his left arm and his will. They say that even in idle conversation he always maneuvers his opponent away from his weak side. From what we know of the wound and the state of the Red Army's medical facilities at the time, it is quite likely that Kerzhentseff was in serious risk of dying. He tasted mortality early, and those of us who are members of that accidental fraternity know that the experience makes a radical change in a man, destroying the simple optimism and leaving him aware that he stands completely alone. It is not uncommon for such men to mend strong.

After his recovery, Kerzhentseff became a junior officer in a unit that administered the repatriation of prisoners. If he ever objected to the policy of universal internment, it has never been reported. The rule was absolute: Anyone, hero or coward, who had fallen into Western hands was considered to be incurably tainted. And Kerzhentseff was efficient in carrying out the work of quarantine. His job was purely administrative, of course; he did not have to look his victims in the eye. But all we know of him suggests that even at this early stage of his career he would not have balked at the sight. It is fair to say that none of the captured Soviets we liberated and nurtured before sending them back ever did us any good. These were very few in number compared with the entire group of returning prisoners of war, and we had some hope that this might shelter them. But when the Soviets face the problem of finding needles in a haystack, they simply burn the whole haystack. It is profligate, but it is effective.

The agency of state security, KGB, was a natural next step for Kerzhentseff. Its headquarters, Lubyanka Prison, is a dungeon of many cells. The KGB permeates every aspect of Soviet society. The rude, leathered thugs on the street corner have their niche in the agency. So do the most advanced technicians. There are units dedicated to the purification of the arts, others to the detection of hidden dissent. But these were not for Kerzhentseff. If I understand him, he stayed away from the censorious parts of the KGB simply because he recognized that nothing so open as performance or so impotent as the furtive complaint was worthy of his talent for deduction and elimination. It was the hidden threat that challenged him, the unspoken treachery, the treason masked by the humiliating deference of the toady. In short, he recognized the pleasures and importance of counterintelligence.

Kerzhentseff first came to our attention because his early work snared a number of our best prospects within the KGB hierarchy. This was all the more striking since they were his superiors. It must have been a delicate game he played because even challenging one's subordinates on grounds of loyalty can be extraordinarily dicey. Toppling the very people you report to takes consummate skill and judgment. His way of operating earned him the nickname Zapadnya, the trap, at least among the defectors who made it safely to Langley before he struck. Zapadnya. It has a number of meanings: the lure of the West, the temptation of the bourgeois, the snapping steel jaws.

Kerzhentseff, you see, was not content to wait for breaches of security to lead him to the weak ones. He had a sixth sense for the qualities that made a man vulnerable, the appetites, the brooding doubts, the very cast of mind that could cause him ultimately to yield. Zapadnya detected these things, smelled them like alcohol on the breath beneath the anise, and went to work. He created artificial situations to test his intuition. He was a scientist of counterintelligence. More than once he tricked one of his own men into thinking he was conducting a sensitive investigation of some Party official's crime, only in the end to reveal, the irony of iron, that the whole investigation was an elaborate entrapment to test his own man's mettle. Many who underwent Kerzhentseff's clever experiments failed. He identified the people we might fruitfully have approached, and he did it long before we even recognized the opportunity. Defectors spoke of him as the principal reason they decided to flee. He drove out those he did not exterminate.

Kerzhentseff was the *dieu trompe* of state security counterintelligence. He never repeated a ruse. He used expectations against his victims, watching them squirming to avoid the last snare, and by the movement he knew precisely where to lay the next. By the time we came to appreciate his style, he had all but destroyed whatever rudimentary network of sources we had pieced together out of the strained friendships of the war years. Zapadnya had triumphed.

Eventually, he turned his attentions outward toward the intelligence agencies of the United States and her allies. He immersed himself in the collection of positive intelligence. Positive. No doubt some would hear that word in a special way. We cannot be sure that his career change represented a decision that the scourge of the weak ones within was complete. But the adherents of the psychoanalytic view of Soviet affairs would probably interpret even this threatening event as evidence of a new maturity and self-confidence in the Soviet political system. Such is their folly that they would see the transfer of Zapadnya as an example of the Soviets' willingness to concern themselves finally with the outside world, to put cultural paranoia, if not out of mind, at least on a lower plane of priority. Oh, yes, they would manage to convince themselves that even in the Soviets' deployment of their most potent human asset directly against us there was evidence of growth. The patient was just acting out, the bloody knuckles a sign of progress.

His techniques, of course, underwent modifications. He was no longer operating in the highly controlled environment of terror. The fear he needed, he had to create. And he could no longer depend on manipulating all the variables. The West, even though he had known it as a child growing up in England, must have presented itself as a complicated puzzle, indeed. Unruly, individualistic, infinitely changeable. A lesser man may have tried to impose the methods of one arena on the circumstances of the other. But not Kerzhentseff. He had suffered a wound, learned his own mortality, known the need to compensate and adapt. I have often said that a man such as Zapadnya would have thrived as well at Langley as he did at Lubyanka.

By the time he came at us, we were prepared for him, or so we thought. He must have sensed that we were. He dealt solely through cutouts. We watched him but, until he contacted

Birch, saw little. At Langley we respected him enough to give
him credit for all our failures: It must have been Zapadnya
who organized the recruitment that compromised our satel-
lites, Zapadnya who found the weakness in our long-lines
transmissions. The direct command and control system we
called Black Body was, of course, attributed to him. He was
just unorthodox enough to push for such a thing even though it
circumvented his own operational control. At Langley a program such as Black Body might have been dismissed out of
hand, not for technical reasons, mind you, but simply because
as a bureaucratic matter it reduced the authority of the very
people who would be responsible for putting it into effect.

Kerzhentseff seemed to thrive, at least until he was reassigned to Tokyo. Some analysts sneered that his exile to Asia
meant that he had finally outsmarted himself. But in fact he
was simply going to the point of our greatest weakness. With
the war on, we were cumbersome in that theater, all exposed.
The Soviets could move freely, but we were at a disadvantage
because the controversy about our cause made us chary of any
misstep. Kerzhentseff had hardly become irrelevant in Tokyo.
He was moving with the opportunities, just as later, when our
domestic purges began in the wake of our foreign defeat, he
was quick to surface in Washington, D.C.

This was why his appearance on Birch's case caused such
excitement. He was the man to deceive. If we could make him
doubt Black Body, his own creation, it would be decisive. And
even though he had a personal interest in seeing his brainchild
succeed, he was strong enough to abandon it in an instant if he
believed it had failed. It did not go unnoticed, however, as
Harper methodically trained Birch to play his role, that we
were not only trying to use Zapadnya's strength against him
but were also developing a locus of power that, despite our
faith in Birch and all our precautions, he might very well attempt to turn against us.

14

Birch was worried about the man, but he was more worried about the machines. Harper had the resources to cope with both problems grandly. No one was clucking his teeth over expenses for this project.

The first step was to build a special new facility at Camp Zama. Technicians in fatigues arrived and took their places on a fictitious new duty roster. They hauled in strange-looking crates used to transport delicate, high-technology equipment and put them in a vacant quonset hut in the security area. Ordinary limitations on access were not enough, so a new fence went up around the installation, tall and topped with barbed wire. The technicians cinched up antennas in the field across from the quonset and made an elaborate business of getting them all at the proper tension and cant. Other antennas were rigged under large tents to obscure them from view.

Birch watched curiously as the devices were wired to conduits in deep trenches. He saw the phantom soldiers move in and out of a barracks within the wire. Harper was too busy to give him an explanation, and it seemed to be a lot of to do just to prepare Birch for the polygraph. Still, the sheer scale of the operation encouraged Birch that whatever it was, they were doing it right.

Though Birch guessed that the new soldiers were civilians in disguise, the imitation of army life was scrupulous to the smallest detail. Someone even thought to put up a sign: 3D JOINT SECURITY DETACHMENT. AUTHORIZED PERSONNEL ONLY. And it was decorated with the ubiquitous cartoon Snoopy, this time wearing an outsized pair of earphones and sitting on the legend: "See no evil. Speak no evil. Hear EVERYTHING."

The rest of the camp seemed to accept the new group as what it pretended to be. The GIs had grown accustomed to the aloofness of the security types. They did not understand it, knew they weren't supposed to, and didn't give a good goddamn anyway. It was somebody else's hassle so long as the new men kept their distance. If the intruders didn't cause the others any extra duty and stayed to themselves, the draftees and lifers didn't care if they were members of some mad, cultish coven.

Birch was assigned as a liaison to the new unit and relieved of his other duties. But for the first several days he was not permitted inside the building that was the center of all the activity. When he finally was invited in, he was surprised by what he saw. The windows had been blackened. The walls and ceiling were covered with a padded material he later learned was designed to absorb electrical emissions. But instead of Mission Control, this room seemed more like a warehouse. Most of the special crates remained unopened, if not empty, in the corners. In the middle of the painted concrete floor stood a big table arrayed with the polygraph machines and some other devices. Around it were folding chairs. And above it a naked bulb hung from the ceiling, its light as harsh and unrelenting as the third degree. It cast everything in stark contrasts, glare or darkness. There was no middle ground. The only other equipment in the room was a TV monitor, a videotape player, and an audio tape deck. The single stuffed armchair that sat next to the table seemed as out of place as a throne in a barn.

"This is it?" said Birch.

"It will be more than adequate," said Harper. "We could

probably have done it with a good deal less, but we wanted to give your friends something to think about."

"My friends wouldn't care," said Birch. "They don't ask me any questions."

"I meant your Soviet friends."

"Have we started already?" asked Birch. "I mean is this part of it now, seeing how I react? The Soviets aren't my friends. But am I supposed to go into that now?"

"It is a bit confusing at times, isn't it, knowing who you are talking to and what they want you to be."

"Just so long as you tell me what to expect, I'll be OK."

"I expect you to be able to make short order of this machine," said Harper. "There are two basic types—the ones that use contact electrodes and the ones that measure the changes in the sound of your voice. The polygraph here is far and away the more precise. It measures various so-called involuntary reactions—body temperature, pulse rate, breath, the electrical conductivity of the skin. The voice device operates on the same basic principle, but it does so at a distance by calibrating changes in the pitch and timbre of your words. It is crude, but you must always assume that they are using it.

"The key to all machines is stress. You might want to think about them as working the way a parent does. I was always amazed when I was a kid at how easily my father could catch me in a fib. Later, of course, I realized that he knew that I expected him to catch me, was afraid that he would. All he had to listen for was the sound of my apprehension. He could always hear the strain, see my eyes avert. Our bodies can give us away simply because we are worried about what they might reveal."

Harper laid his hand on the shiny chrome device and switched it on. A strip of paper inched through it under a set of pens attached to an array of gangly limbs like a spider's. The lines it drew were flat and parallel but for an occasional random blip.

"Just remember, it does not read your mind," said Harper. "It only reads your body. And then only the precise measure of your fear."

"What about drugs?" asked Birch.

"They don't trust them. Neither do we. People react so differently. You might get the truth. You might get a deeply held fantasy. All you can be sure of is the high. We can try them later if you like."

"No, thanks."

"If you can beat the machines, you can beat the chemicals."

"Willpower," said Birch.

"I like to put it somewhat differently. You have to make your cover story as much a part of your flesh as the muscle of your arm. It isn't as difficult as it sounds. Sit down and we'll have a start at it."

Much later Birch had reason to complain that no one had explained to him the risk of what he was about to learn. If he had known, he might not have agreed to sever the one verifiable link to the truth. It may seem unfair of us to doubt him after having taught him how to frustrate all the techniques of corroboration. But the Agency is not chartered to be equitable. We do not give Miranda warnings. We have learned to live with the sad truth that the more capable we make an asset the more dangerous and endangered he becomes.

At the time, though, Birch did not worry about anything but the challenge at hand. The first order of business was to prepare him to explain his travels. Harper said that the KGB's speed in taking over the case after disposing of the Slav raised the possibility that it had been observing Birch all along during the earlier stages. There were blocks of time to account for: the visits to the safe house, the trips to drop sites. Harper led him through the movements before and after those rendezvous, then together they created stories that as closely as possible matched the real map of Birch's wanderings. As they practiced, Birch could almost hear the voices of the Japanese ladies in the imaginary shops he browsed during the time he was in fact at the safe house. He could feel the pressure welling up in his bladder before he leaped the cemetery wall to relieve himself in desecrating privacy. It was not so difficult to replace the one memory with the other. The recent past had been so fabulous and strange that it seemed like a dream anyway; he could

not escape the vague and distant sense that what he was
doing was as unreal as what he was trying to pretend.

They went over the locations Birch had given the Slav for
every document he had provided. Harper consulted a detailed
summary he had prepared of everything the Soviets had been
told. But it was not really necessary. Birch had every piece of
paper neatly filed in his mind, every tumbler of the combination lock on the imaginary safe ready with its authoritative
click.

Next Harper provided a sketchy reason for the presence of
the new unit. Birch was to know no more for now. His rank
would not privilege him to any details at the moment. There
was a drawing of the layout inside the quonset hut which Birch
quickly committed to memory. It was considerably more elaborate than the real setting that surrounded him, more like what
he had expected to find. He liked it better the way he was to
tell it than the way it was, the fancy, imaginary arrays of dials
and switches and flashing lights, the silent, spinning computer
tapes, the hushed, ethereal whisper of the wires. Harper gave
him a quick look at pictures of some electronic gear, and Birch
was to pass on anything he could remember from that brief
glimpse.

Finally, they moved on to the machine. A technician
strapped Birch uncomfortably about the chest with an elastic
belt to measure the depth and rate of his breathing. He daubed
Birch's wrists and forehead with an icy, conductive gel and
wired him up to the electrodes. When it was done, Birch felt
bound, hemmed in. What's more, the ritual of it, the vest-
ments, the touch of holy oil, it was a kind of oath. He wondered whether it wasn't all simply meant to impress him, to put
him on edge, all part of that business about the fear of fear.
Well, he was not going to let it get to him. He took a deep
breath against the strap and shook his head, the wires dangling
like a cowlick.

"Comfortable?" said Harper.

"Real nice."

"The first step in fluttering is to get a calibration, to see how
you normally react," said Harper. "The device the Soviets use

might look a little different than this one, but it won't be any better. This is the state of the art, made in Japan."

"I guess I will be too," said Birch.

The technician flipped on a switch, and the tape began to scroll out under the swinging pens. Birch watched the graph draw itself. He took a deep, tight breath, and the fragile needles careened wildly. It was an odd sensation, like looking at yourself in the mirror and wondering what those eyes, which you could never really meet, might reveal.

"You must answer these questions by a simple yes or no," said Harper. "Anything more discursive will throw off the measurements. Do you understand?"

A. Yes.

Q. Is your name Jerry Birch?

A. Yes.

Q. Were you born in September of 1946?

A. Yes.

Q. Are you married?

A. Yes.

Q. Do you have any children?

A. No.

Birch felt an odd tug in the belly. He was acutely aware of his own reactions now, the complex feelings this simple statement summoned up in him. In other circumstances he could simply have said it without giving it any special thought. But now he dwelt on it. How much Donna wanted a baby. How they had tried. How he had once used their failure to hurt her. It was as if the electrodes taped to his skin were looped back into his own nervous system, feeding back the signals, amplifying every emotion.

"Now I want you to notice something, Birch," said Harper. "See how the pattern fluctuated?"

"I figured that it would."

"You were, I assume, telling the truth. But something made you respond."

"We've always wanted kids," said Birch.

"My point is that such variations are normal and by no means indicative of a lie. The operator will notice the pattern

on innocuous questions and that will provide him a bench-
mark. Each individual has his own unique polygraph finger-
print. This becomes the standard of measurement. And if you
are clever, you will make sure that the graph on these questions is not too smooth so that later it will be easier to keep
within what appears to be your normal range. This is our strategy, and in later sessions we will work on it. But for now, simply be natural. We'll continue."

"Natural," said Birch. "It's like telling a man not to think of
a green elephant."

"I was meaning to mention that," said Harper. Then he
paused while Birch's curve relaxed and they were ready to go
again.

Q. Are you a staff sergeant in the United States Army?
A. Yes.
Q. Do you draw the regular pay and allowances?
A. Yes.

"I would have called our contributions a little bit irregular,"
said Harper. "Wouldn't you?"

"I suppose."

"The question had a hidden edge. Did you feel it?"

"Not particularly."

"How does it look?" Harper asked the technician.

"Maybe he didn't understand the question."

"I understood it," said Birch.

"Well, it didn't seem to bother you," said the technician.

Q. Do you know a man named Nowicki?
A. Yes.
Q. Have you provided him with classified information?

Some involuntary hand made a fist in Birch's belly. He could
feel his breath growing short. The electrodes on his forehead
seemed to be sending out a signal to his brain that made him
queasy and confused.

"Still troubles you, doesn't it?" said Harper.

"Am I supposed to answer yes or no?"

"It is good that you don't take the matter lightly," said
Harper. "It makes a very convincing showing on the tape."

Q. On July 17 did you go to the cemetery in Rappongi?

A. Yes.

Harper hadn't told him whether he was supposed to try out his story, but Birch was curious to see how easily it would come out. If anyone from the KGB had seen him stuffing the message in the little shrine, he was a goner. The rest, as he now remembered it, was simple to explain. He had been very nervous that day. It was all so mysterious and new, this business arrangement with Nowicki. The itch in his groin had been real, the guilty pressure he used to feel when he did some foolish thing on a dare. The men who had been in combat used to joke about it. The first rule of battle, they said, was to hold it all in.

Q. Did you climb a wall to enter the cemetery?

A. Yes.

Q. Had you ever been in the area before?

A. Yes.

Q. Did you know there was an open gate?

A. Yes.

The way to explain it had been Birch's own idea. He had made it up out of fact, slightly refashioned to need. And now this was precisely the way he remembered it.

Q. Did you climb the wall looking for a private place to urinate?

A. Yes.

Q. While you were in the cemetery did you leave a message for anyone?

A. I just had to piss.

"Please answer simply yes or no," said Harper.

Q. While you were in the cemetery did you leave a message?

A. No.

Q. Did you arrange with anyone to be at the cemetery?

A. No.

Q. Did you urinate before climbing back over the wall?

A. Yes.

"Excellent, Birch," said Harper. "I doubt that even our expert here will be able to find anything out of order in that performance."

"No," said the technician. "It looks pretty good. Of course, the subject was not at real risk."

"Sure, sure," said Harper. "That comes later. You have to admit, though, that he's off to a good start."

"I could make something out of the heartbeat here," said the technician, pulling the long folds of paper through his fingers. "Do you see the slight aggravation?"

"It could have been anything," said Harper. "You couldn't read anything from that alone."

"Ambiguous," said the technician. "By the way, did you really take a leak?"

"Sure I did," said Birch. "Just like I said. Do I get a gold star?"

"Keep it up and I'll mint a medal for you," said Harper, "crossed fingers on a field of azure."

"I guess I'd have to pretend I never received it," said Birch. "Of course."

Q. Have you told anyone of your meetings with Nowicki?
A. Yes.
Q. Have you told your wife?
A. Yes.
Q. Have you told your superiors?
A. No.
Q. Have you discussed this matter with any U.S. agency?
A. No.
Q. Did you have reason to believe Nowicki was a Soviet agent?
A. Yes.
Q. Did you have reason to believe this before you spoke to Kerzhentseff?
A. Yes.
Q. Did you deny to Kerzhentseff that you had reason to believe it?
A. Yes.
Q. Did you lie because you feared the knowledge put you in legal jeopardy?
A. Yes.

Q. Are you lying now?

A smile spread across Birch's face as he remembered the math puzzlers he enjoyed in the science magazines. Mind-twisters. They were a kind of joke. There was no answer to them, none in all the universe.

A. Absolutely.

"You are pleased with yourself," said Harper.

"Not really," said Birch. "It's just that if you say you are telling the truth you are lying. And if you say you are lying, then you are telling the truth."

"Epimenides' paradox," said Harper. "It is a comfort in our line of work."

"Who was Epimenides?"

"To tell the truth, I don't know," said Harper. "You should be happy with yourself. You are testing out strong."

"It's funny when you think about it. I mean, we're grown men."

"Playing with expensive toys."

"Playing tricks on ourselves," said Birch.

"It's better than guns."

"That's what Kerzhentseff said."

"He's a liar," said Harper.

"I'm not sure I'm supposed to find that funny," said Birch.

"Laughter doesn't compromise you. The machines can't distinguish it from fear. The needles jump, and it's the funniest thing in the world. Perfectly ambiguous, whatever they may suspect. Laughing can save your life."

"Somehow I can't picture Kerzhentseff hiding that way," said Birch.

"He might have had a laugh over the fate of your friend Nowicki."

"Yes," said Birch. "Maybe about that."

Q. Did you meet regularly with Nowicki?

A. Yes.

Q. Did you meet for dinner?

A. Yes.

Q. Did you ever meet in a park?

A. No.

Q. Did you ever meet in a brothel?

A. Yes.

Q. Have you ever made love to a woman other than your wife?

A. No.

"Wait a minute," said Birch. "Turn that thing off. What does this have to do with anything? You know what I did that night and why I did it. They know too. Even Kerzhentseff promised to stay out of this stuff."

"And you believed him?" said Harper. "I'm going into this for a reason. I want to give you a little workout. You have to feel it. You know what the coaches say."

"What's that?"

"No gain without pain," Harper said, too smugly.

"I thought you were just calibrating the machine."

"That's why we have to go to the stress points, to see how you react."

"Now you've seen."

"Yes," said Harper. "And it disturbs me. You must not show any weakness. They'll drive a wedge right into it."

"I want to protect my wife. I'm old-fashioned. I thought we had that settled. I thought we agreed that was all right."

"We aren't threatening her. That's all over. I was against it from the very beginning."

"You didn't seem to be."

"We all just do what we're told."

"But you told them their plan was wrong?"

"Persistently," said Harper. "They weren't listening. But now they are, I assure you. There will be no more of that from us. But I can't say what the Russians will do. You have to prepare yourself. We have to get a picture of your response so that if it needs touching up we can do so. Now, can we go on?"

Q. Did you meet your wife while you were in high school?

A. Yes.

The pens traced smooth, easy curves on the paper as it inched under them. Birch was calm again, satisfied that he was safe with Harper, that Harper understood. You had to trust somebody somewhere. You couldn't put everything on the questions and the answers. There was a factor beyond truth, or at least separate from it, that you had to hang onto. Birch found the quality in Harper, and now he felt it had been reaffirmed. He assumed, because this was an essential part of the intuition of faith, that Harper believed as strongly—and without regard to proof and disproof—about him.

Q. Was she in your graduating class?

A. Yes.

Q. Did you marry shortly after you graduated?

A. Yes.

Q. Had you ever had sexual intercourse before your marriage?

A. No.

Q. Had your wife ever had sexual intercourse before your marriage?

A. No.

Q. Since your marriage have you ever had sexual intercourse with anyone other than your wife?

A. No.

Q. Have you reason to believe that since your marriage your wife has had sexual intercourse with anyone other than you?

A. No.

Birch was mesmerized by the droning questions. They touched on the most personal matters, the deepest issues of the flesh, and yet like questions of faith and spirit in church, they sagged into the dull, reassuring litany. I believe in God the Father almighty, maker of heaven and earth . . . Even the most important things could be spoken aloud so long as you had no doubt. I believe in the forgiveness of sins . . . He had nothing to be ashamed of. The words did not gag him. They came out as smoothly and automatically as the slow beating of his heart.

Q. Have you ever had occasion to desire sexual intercourse with someone other than your wife?

A. . . . .

He did not need to look over at the chart to know he was sending the pens into a jagged outburst. He could feel the sweat beading at his pores, his breath going hollow.

A. Yes. No. No, I haven't.

"Which is it, Birch?" said Harper. "You decide."

"It isn't easy."

Q. Have you ever had a sexual encounter with anyone other than your wife?

A. No. Yes . . . You've changed the question.

"The Russians won't make it easy on you, Birch."

"Give me a few seconds."

"We must go on."

Birch breathed deeply and opened and closed his hands a few times to relieve the tension there.

Q. Have you ever had sexual intercourse with anyone other than your wife?

A. No.

Q. Since your marriage, have you ever been tempted to have sexual intercourse with anyone other than your wife?

A. Yes.

"It's curious," said Harper, as easily as if he were referring to some quirk in the weather. "Look here. You show a pattern characteristic of dissembling, no matter how you answer that question. Isn't that strange?"

"It's your damned machine."

"I don't think so."

"Look," said Birch, "I've told you what happened with the girl. You tell me, did I want her?"

"You'd have been an uncommon man not to."

"What is that supposed to mean?"

"Everyone is tempted from time to time. Most with considerably less provocation. She's very pretty. I've seen her picture."

"You made pictures of it. Oh, my God."

"No. No. Not at all. We had to check her out afterward. Just to be sure. She turned out to be nothing but what she seemed."

"A harlot," said Birch.

"Another view is that she is a small-town girl come to the city to support her family. Would you like the whole life story?"

"I don't give a damn. It will just mix me up even more. When you ask whether I was tempted, I can't answer either way. It was wrong, no matter which. The truth is that it was everything at once. Do you understand that?"

"It means you are free to say whatever you please," said Harper.

"I guess I am."

"In some sense, there is truth in every statement you might make, if only truth to the commitments you have made. That frees you. You can choose which truth to tell."

"You make it sound easy."

"I'd say you were making it look easy, Birch. I'm just here to give you a few pointers. You're going to be a quick study. One thing to remember is that if you feel yourself getting into trouble, give them more details. Don't just answer yes or no. Pour it out. It will confuse the measurement and buy you some time. You might even try coughing."

"But when they demand an answer . . ."

"Then give it to them quick and sharp. Don't even think about it. Let your training take over and speak for you. By the time we have finished here, I'm sure you will be confident that you are fully prepared. Now, just one more line of questioning, if you don't mind."

"Sure," said Birch.

Q. Have you ever been arrested?

A. No.

Q. Have you ever been employed in law enforcement?

A. No.

Q. Have you ever been charged with perjury?

A. No.

Q. Have you ever undergone a polygraph examination?

A. No.

Simultaneously, the two of them glanced over at the unfurling strip of paper and saw the gentle sine curves of his rhythms rolling easily up and down. Then Birch laughed out loud and the needles spiked. And, though he had nothing to be afraid of, Harper laughed too.

15

A wedge of geese paddled past on the inky water of the moat. Three neatly intersecting wakes, like wave diagrams in the electronics manuals, were the only flaw in the sapphire surface. The high, curving stone wall across the water doubled its size in reflection. And beyond it rose the imperial palace, fortified against other centuries' dangers.

The sun sizzled in the fumy haze that in this place seemed almost natural, and Birch wondered why Kerzhentseff would choose such a restful spot for him. At the quonset hut they had warned that the Soviets would try to jangle his nerves, put him on the defensive the better to trap him in a lie. Birch had trained there hour after hour, wired up like a patient in intensive care. In time, the straps and metal contacts felt as familiar as the squeeze of his collar button or the sharp ridge of his low quarters across his shins. The ugly questions and insinuations became little more troubling than the mindless subordination of a military salute and salutation. It was all a matter of drill.

The answers about the past came easily to him now. Even his encounter with Toshi was clear in his mind. He had not desired her. Not in the least. He had been faithful in his heart. The machine did not trip him up because this had become for him the truth. He did not even have to try in order to make the needles trace gentle, rolling waves any more than the ducks had to bark at one another to stay in formation. It came naturally to him; to try was to fail.

A group of tourists moved up next to him at the edge of the moat, chattering in some European tongue. One of them—an older woman whose wattles hung out over an expanse of bright, flowered fabric—began arranging the others for a group picture along the low wrought-iron fence edging the water. Some distance away a young Japanese stood meditating.

"Here, let me," he said, stepping up to the woman and with an awkward charade offering to snap the shot himself so that the whole party could be in at once. The stout lady agreed and explained to him excitedly the operation of the camera in words not one of which Birch understood. He nodded, pointing to the shutter button. She said something several times over as she retreated to the group. Yes or thank you, he hoped. She bounced her way into the center of the line and put on a celluloid smile.

If it was to be a secret rendezvous, he did not want a record of it lurking in somebody's scrapbook. If Kerzhentseff had eyes watching him, he would be reassured by Birch's quick thinking. Through the viewfinder, Birch watched them stirring themselves to order. This one straightened her skirt. That one undid the second button on his shirt in order to be more rakish. Birch stepped back two paces, his eye still fixed to the lens, so as to get them all in—hats to sandals—with a little bit of the palace grounds behind to give the picture its motive. All smiled except for the fat man, who looked incapable. In the distance, around the bend in the moat, the young Japanese moved away. Birch held the group in frozen cheer until he was out of sight, then he snapped the shutter.

"Another," he said, waving them back into place, "just in case."

When it was taken, this time from a low position on one knee to make them seem taller, he handed the camera back to the woman and declined the coins she tried to press into his hand.

"I'm free of charge," he said, realizing immediately that he should have accepted. It was always easier to forget a hired man than a courtesy. But it was too late. The group had ambled off to new backdrops.

When Birch got back to the railing he saw what he'd been waiting for. It was there, toed into the gravel where he had been told to wait, the international symbol of peace, but this one squared instead of circled so as not to be mistaken. The young Japanese was a hundred yards past him now, moving away briskly and not looking back.

The signal was to be timed. This was the order. Move when you see it. Do not give a countersignal. Do not hesitate. Birch obliterated the lines in the stones with his foot and started across a wide lawn. The young Japanese had watched him for nearly an hour before making the move. They either did not trust him or they were going out of their way to protect him from detection. Or both. When he reached the street he checked his watch. He had six minutes to reach the commuter train station, and he figured he had already used up two of them. The traffic was too heavy to cross against the light. He waited impatiently. Then at a brisk walk he cut through the lobby of a building to confuse his imagined pursuers. Everything according to the plan Kerzhentseff had laid out in the car. He reached the tracks and followed them to his destination. It was crowded even at midmorning. Birch queued up to buy his passage. When he reached the window and put down his fare, there was still no sign of the train. He was not to commit himself to one platform or the other until the last moment. So he loitered near the subway map on the concrete wall until the train rumbled in and came to a stop. Then he hurried to the stairs and up them. Like a cog in a precision machine, he slipped through the pneumatic doors just as they began to close.

There were empty seats on the car, but he chose to stand as the train jolted forward. He was the last man in. It had worked perfectly. He was, in a sense, safe.

The car jostled from side to side, and Birch planted his feet wide to keep his balance. His hand tightened on a chrome rod near the door as he leaned slightly forward to see where the ride was taking him. Buildings rattled past, and he gazed into third-story showrooms and offices, a voyeur, wondering who might be looking back.

"They expect you to be afraid of being caught," Harper had told him. "They think you are an amateur. They assume you will worry about being seen."

"I will," Birch had said. And now, without trying, he did.

At the fourth stop, he slipped off the car and waited on the platform until all the others had made their exit. Then he checked his watch again. Already elapsed: seventeen minutes. Right on time, he bounded down the steps and to the street. A car was just pulling up, its door swinging open to him. He slid in without breaking stride and pulled the door shut as the driver turned into traffic without a word.

There were no signs or countersigns for this stage of the plan. It was a strange procedure, silent and vulnerable. Suddenly he wondered whether he might not have made a mistake. Was he in the right car? What if he had stumbled into somebody else's story? He caught himself in a nervous chill and made himself laugh.

The driver watched this attentively in the rearview mirror. Birch met and fixed the hard eyes there to prove a point to himself. They finally flickered away.

Nobody was watching Birch now but Birch himself. He was conscious of his own actions, like a moviegoer watching a character on the screen. Grainy, washed-out colors on the shadowed streets. Sky going overcast. Bright, hazy sun in retreat. The threat of rain. He heard footsteps, troubled breath. He identified with both the cinematic hunter and the quarry, for they were one. He was immersed in the story, but still it was much more realistic than real.

They said this was what happened to you in combat. They said that some saving mechanism in the brain took you outside of yourself and, watching, told you what to do. They said in battle you became John Wayne or you died. He had disbelieved these things, the stories about how, suddenly, in the crackle and thunder of a firefight you could pause for a moment, belly hugging the earth, fingers clutching at the grass like it was life, and find the whole thing unbelievably funny. But he didn't disbelieve anymore. He was James Mason, riding to an unknown assignation to fulfill the hidden imperatives of the script. The old print jumped and flickered before his eyes. Someone was whistling a haunted theme. The hell with the driver. Birch laughed out loud. And then he settled back into the seat, the tune rising above the hum of the engine, and he eased his uncharted readings into long, graceful curves.

Birch did not recognize the streets they were following. In his distraction, he had not paid attention to the route, and by the time he came back to his obligations, he found himself in an unfamiliar part of it.

When the car slowed to a stop at the curb, Birch did not move. He was waiting for instructions. The practiced plan was at an end, and he was at the disposal of someone yet to show himself. The door swung open automatically, like a taxi's, and the driver motioned toward it with a silent shake of his head.

"What next?" Birch asked as he slid out.

The driver did not answer, did not even turn.

"I mean, where am I supposed to go?"

The door shut and the car pulled away. The street was empty, desolate. Warehouses and loading docks crowded either side, as anonymous as freight. He stood alone on the sidewalk and tried to remember the exact words.

"We will know if you have been followed," Kerzhentseff had said. "If we make the final approach you will know that it is safe. The sign to you is 'travel plans.' You will ask directions to the Shinto shrine."

But which way to turn? There was no way to divine it from the instructions, no one on the street to give him guidance. He walked to the corner where there were some stores. He stopped in front of one of them and watched the moving display of appliances.

Something had gone wrong. He had done something that put them on alert. Harper had promised not to follow him. But how good was a promise? Birch panicked where he stood, defenseless and exposed. He fought the urge to slip inside the store, just to be out of sight. But that was no escape. It was a cul-de-sac. These people did not screw around. They took you to a place where you didn't know where to run. This was the part of the movie where the shot rang out, the man fell.

"Travel plans?"

The voice caught him from his blind side. He spun around, too quickly, showing his fear. He was in the wrong movie, playing the victim's part. And everyone saw the danger but him.

The man with the voice looked nervously over Birch's shoulder at the empty street. He was a thick, small man with red, thinning hair. His eyes were hidden behind mirrored glasses that gave him the appearance of a reptile. Beneath his molting suitcoat he could have secreted a gun.

"Travel plans?" he insisted.

"Yes," said Birch. "In fact I . . . well, I'm trying to find a shrine."

"A Shinto shrine?"

Birch tried to place the strange sibilant accent, but it came from nowhere he knew.

"That's it. Shinto."

"I think I can help you. Come this way."

He took Birch by the elbow, his fingers boring in on the nerve like punishment.

"That hurts a little," said Birch. But the man held fast, leading him into an alley where even James Mason might have balked. Birch had run out of script. This is where they slip in the blade. Here is where the pistol glides smoothly out of his holster and makes one dull thump before the screen goes black. And when they find the body they can only admire the professionalism of the job.

The man opened a scratched-up service door and nudged Birch inside. Birch stumbled and turned in time to see the door closing on the man's squat figure. Then the door crashed shut, and the darkness was more than an absence of light.

"You seem to be a little confused, Mr. Birch," said a voice, whispering. It was Kerzhentseff.

Birch groped toward the sound and banged into something low and hard against his knees.

"Maybe you had better be still until your eyes grow accustomed. I'm afraid the way is strewn with obstacles."

His fingers felt the thing he had struck. It seemed to be a heavy wooden box. The slats were rough and slivered against his skin.

"Please sit down, Mr. Birch. You'll have to excuse the uncomfortable surroundings. We do take some pains to be obscure. A light would only attract attention."

Birch managed to make out the shadowed figure of the Russian sitting a few yards away. The room appeared to be a large storage area of some sort. There were bulky packing crates stacked here and there, but otherwise it was empty from rafters to concrete floor. The question was not whether it resembled the quonset hut at Camp Zama but whether it was intended to.

"I think you will get used to our precautions," said Kerzhentseff. "At first some find them a bit disconcerting. But in time you will find them less mysterious. And I'm sure you will see the advantage in being careful."

"When the driver left me off alone, I was afraid something had gone wrong," said Birch.

Light seeped into the room through one distant, shaded window. It was only enough to give the outline of things. He could not see Kerzhentseff's expression. And, not being a machine, he read nothing in the sound of his voice.

"What made you think there was a probelm?" said Kerzhentseff. It was a primitive trap. He could not have been serious about it.

"I'm new to this," said Birch, relaxing now. "I didn't have any directions."

"You stayed put until you got your orders," said Kerzhentseff. "That was proper. Never do anything until you are told."

"Hurry up and wait," Birch said. "Like the Army."

"Precisely," said Kerzhentseff, and there was no humor in it at all. "We will begin alone. There are preliminaries that must be dealt with. Later, perhaps, we will be joined by others with the technological background to appreciate what you are in a position to provide."

"I'll talk to anybody," said Birch.

"You will talk to no one except in my presence."

"That's what I meant."

"Of course," said Kerzhentseff.

"To tell the truth, I was feeling as silly as scared before," said Birch. "I even wondered whether I had gotten into the wrong car."

"There is nothing silly in this," said the motionless figure across from him, closing off the refuge of laughter. "If you think so, it is only because you have seen too many melodramatic films."

"You must have read my mind," said Birch. "I was thinking about that, about James Mason, you know. Maybe you've never seen any of his movies."

"The trouble with films is that they always work themselves out in the end," said Kerzhentseff. "The mystery is in what is held back, not in what is given. It is little challenge to discover what the director has known all along. I find the films and the fictions rather too simple, rather too . . . optimistic. If I were you I would envy James Mason."

"Oh, I do," said Birch. "He is a very good actor."

"And so are you, Mr. Birch. So are you."

In the silence that followed, Birch fastened his attention on the exhalation of some sort of fan and the muffled sound of traffic outside. A muscle in his arm began to twitch. If he were wired, he knew that the pens would be spiking now. But all they could monitor was his voice; he was safe until he spoke.

"You are not as innocent as you like to seem," said Kerzhentseff. "If we are to go on, you must not continue to treat me as a fool. You can begin by telling me of your contacts with counterintelligence."

"I was interviewed for my clearances," said Birch. "That was two years ago."

Dangerous as Kerzhentseff was, Birch felt sure he would be

all right so long as he stayed with what he knew, and this was part of it.

"This is not what I mean, Mr. Birch," said the Soviet. "Please try to understand that I do not intend to indulge in simplicities with you. If nothing else, remember my capacity to do you harm. We know they had Nowicki under surveillance and caught you up in it. What did you tell them, Mr. Birch?"

He will bait his trap with whatever comes to hand, Harper had said. He will pretend to know facts that contradict you. But do not bend to them. Never bend. If he has seen anything fundamental that we have not prepared for, the game is over anyway. You must assume that he does not know the ultimate.

"I didn't talk to anyone," Birch said, "not even my wife. She thought he worked for the electronics firm."

Birch was not perfectly easy in his answer, but the twitching muscle had stopped and his voice was as calm as the circumstances called for.

"We know how they operate, Mr. Birch. When they see an unauthorized contact, they make inquiries. There is no sense in lying to me."

"No one has asked me anything about Nowicki," said Birch. "My God, are they on to me? Is that what you are saying? Why didn't you warn me about it? You're going to destroy me."

He let all his anxiety drain into these words, pushing the hidden needles to jagged peaks of stress. It was cathartic, and afterward his body felt free of the poison. His fear was credible and corroborating.

"You will be the first to know your peril," said Kerzhentseff. "What have the Americans asked you?"

"Nothing," said Birch. "I had no idea they suspected anything. I've even been given a new assignment."

"Perhaps to get you out of the way."

"I don't think so," said Birch. "They seem to take it very seriously."

"We will come back to that. But first, I must say that it is implausible that you were never noticed with Nowicki. The restaurants. The whores. You were notorious."

"If my family finds out about this it will kill them," said Birch.

"Don't hide behind your family, Mr. Birch. You are alone now. And I do not believe you."

"What am I supposed to say? You tell me that I've already been found out. You say that Nowicki wasn't careful. And then you tell me that you don't trust me. You don't trust me. It's funny, don't you think. Are you laughing? I can't see your face."

"Nobody thinks of it as trust," said Kerzhentseff.

"I'm alone, all right. I'm out at the end of the limb and you are laughing at me. What did you bring me here for anyway?"

"Perhaps I was mistaken," said Kerzhentseff, as if it were a very small thing. "Nowicki was, in all events, undistinguished. Perhaps the Americans paid him no mind."

"Perhaps this. Perhaps that," said Birch. "I've got to know."

"And so do we, Mr. Birch. So do we."

"I'm not going to dig myself a deeper hole," said Birch, standing abruptly as if to leave. He was pleased with himself now. He was on the offensive, and he felt the rightness of his approach in every sinew. Kerzhentseff was now in the position of having to reassure him, to bring him back into the fold. Birch was relaxed, and he felt in command. He was not the first person to underestimate Kerzhentseff.

"Sit down, Birch," Kerzhentseff said. "I don't think you fully appreciate the situation. We have nothing to lose. You have everything. Nowicki is no longer useful to us. You have no means of threatening me, you see. You have nothing to sell the Americans. But if you are telling me the truth, if the Americans have not questioned you and have no reason to suspect your loyalty, then the moment you decide to put an end to our association, we will have nothing left to do but to compromise you. And if you are lying, we will find you out. I assure you, we do not take such a matter lightly.

"So, as you see, it is in your interest to persuade me—your deepest interest. And if you do, if you show me that you can be useful, then perhaps there are things I can do to make sure that the Americans will not find out."

When the situation becomes difficult, Harper had said, slow down the pace. Take your time. Don't think about it; your stories are set and must not be changed. But drag the process out just to give yourself time to breathe. Birch waited, breathed, tried to get the measure of the man half-seen across the room. But there were no meters that would read that voice. If you wired him up, the chart would be flat as a dead man's.

"What can I do?" Birch finally asked. His submission was, in one sense, total. He gave Kerzhentseff the day. This was not a man you could fight.

"As a beginning, you can tell me about this new assignment of yours. Perhaps that will give me some idea."

Birch told him of the arrival of the new unit and the equipment that came with it. Kerzhentseff wanted details. Dates, numbers, how many of each rank. He turned Birch away from the central elements that Harper wanted to convey, and Birch was too weak to force him back to the point. It was as if he had been there days without sleep, his answers were so heavily burdened. He was worn down, beaten, and he let the Russian lead him wherever he wanted. Yet, when all his defenses were stripped away, what was left was the story: the unit's duty roster, the glimpse of the gear. To try was to fail, and in his fatigue and submission, Birch found himself getting the story out. The routine. How many worked in the hut at any given time. The names sewn above their breast pockets.

"And were you told what the mission might be?" asked Kerzhentseff.

"No, sir. I was just to be a liaison, help them to get settled, grease the way when they needed something. The only time I got inside the quonset hut was by mistake. I followed the lieutenant in and was back outside within a minute."

"Were there other special security measures?"

"Yes, sir."

"What were they, Birch?"

"A special pass to get through the gate. Colonel Robertson and I are the only ones from our outfit to have them."

"What else?"

"A briefing, sir."

"An interrogation, you mean."

"Sure. Maybe they asked me some questions, but just the usual ones about was I a member of certain groups. I had to sign some forms."

"There was a new clearance then?"

"Yes, sir."

"And what did they call it, Birch?"

"The name of the clearance was classified top secret. I'd never heard it before."

"You are hesitating."

"They called it Black Body. Does that mean anything to you?"

"It means that you tried to deceive me."

Only in the sense that had been prearranged, only with the purpose of being discovered. "I didn't lie," said Birch.

"You have spoken to American counterintelligence," said the Russian. "And not two years ago. That is disturbing to me."

"They didn't ask about Nowicki."

"That isn't the point. I am not asking a lawyer's questions. I do not expect you to cut your answers thin. Were the people who interrogated you in uniform?"

"Yes, sir. But they were civilians."

"How do you know that?"

"I could just tell. The way they stood. The way they spoke to me. The slouch when they sat in the chairs."

"Sat?"

"One of them rocked back and forth on the swivel, just rocked back and forth listening. He wasn't Army."

"Very observant," said Kerzhentseff, but in a tone that made it clear he was not impressed.

"I was scared when you said they were investigating me," said Birch. "I didn't want to believe that they were."

"You tried to hide it from me."

"From myself, sir," said Birch. "They just asked the usual questions, made me sign the papers. There was nothing to it. It didn't mean anything, did it?"

"Behavior," said Kerzhentseff. "That is what we want you to

report. Facts. The surfaces of things. The rest is conjecture, and your assumptions are not in the least interesting."

"Yes, sir," said Birch. The point of maximum peril, Harper had called it, the point where you have nothing more to yield.

"I want you to come here for a moment. Come."

Birch groped his way forward past the crates. He moved slowly, carefully. And when the dazing flash of light came, it caught him bent and unready. His hands rose instinctively to his face and he heard himself give a tiny, useless cry against the coming pain. But then there was silence.

"It's only a tensor lamp," said Kerzhentseff. "It won't harm you."

Birch looked out from behind his crossed arms and saw the small man wearing what might have been mistaken for a smile. He sat in a deck chair before a low, bunged-up table. The light shone down on its scarred surface and back up again low against his face, hollowing out the Russian's eyes and sharpening the edges of his features like the passing of years. It brought out an Asian aspect to the man, Mongol, something about the way the skin was, sallow, and an ominous shape in the eyes. They are not like us, Birch thought. They are from a different world, and the difference is written in the bones of the skull.

"What would you make of this behavior?" Kerzhentseff said.

He slowly spread five photographs on the table. Birch recognized himself, blurred and grainy. The backward glance. His awkward jogging. The leap over the cemetery wall.

"Odd, wouldn't you say?" said Kerzhentseff.

Birch laughed uneasily, as he was trained. But the chill grip of Kerzhentseff's one good hand closing on his wrist caught him short.

"Have a closer look," the Russian said, his fingers digging into Birch's flesh and dragging him farther into the light.

"I remember that day," said Birch. "I was out walking."

"A man does not forget the experience of trying to hide."

"I was embarrassed," said Birch.

"You thought you could evade them. Then you lost your nerve."

"I had to piss," said Birch. "All that coffee. I imagine you have a picture of that too. Birch relieving himself." He could feel the pressure in his bladder. The guilty sense of desecration. He would not have been surprised if they showed him the photograph of what never occurred—poor in quality, his hands a smudge at his fly.

"You had reason to believe the Americans were following you that day," said Kerzhentseff. This was not a question but a proposition. And, yes, Birch could remember the feeling of being watched, the strange, exciting sensation that someone was about to discover him at his unsavory business. And it had registered itself in his groin, a pressure there, a daring thrill welling up.

"Had you noticed something?" asked Kerzhentseff. "Had the interrogation come close to the mark?"

"This was before," said Birch. "Before Black Body."

"Then why were you afraid?"

"I don't know," said Birch. "There wasn't any special reason. Nothing I could point to and say that was it. But I was going to meet with Nowicki. He had called. I was nervous about it, what he would want next. I'd never broken the law before."

"Your honor," said Kerzhentseff.

"I don't know what you would call it."

"I would call it fear, Birch. You are a coward."

"Maybe I am," he said. And again he knew that the calculated admission was also true. He was afraid of combat, afraid of being killed. He was loyal to his wife, not only because it came naturally to him, but also because he was afraid to lose her. "I've never said I was brave."

"We appreciate cowards, Birch. They make good servants."

The humiliation of it was real. Birch suffered it as a punishment that he deserved.

"If the Americans do have doubts about you, we will expect you to be careful. You must assume that they do. That comes easily to you, doesn't it?"

"I guess it does," said Birch.

"And we will also have to take certain steps to preserve the integrity of our relationship to you."

"You will protect me?"

"This is part of the compensation we offer. But you must leave that to us. We will make sure that whatever traces you and Nowicki left are erased. For your part, you will simply proceed with your work and remember every detail you can about this operation, this Black Body. We may have other assignments as well."

"What traces are there to erase?" asked Birch.

"Unfortunate recollections. It is none of your concern. You must only rely on our experience with this sort of thing."

"I'm sorry I didn't tell you everything right away. I don't know what is expected of me."

"We have no illusions, Birch. We expected no more than you have shown."

16

By the time Harper had finished writing his cable, the maids were already gone. They had darted and bustled around his desk earlier in the evening, not daring to speak. But he had barely noticed them. The corridors were empty as he locked his desk and put the handwritten document into his safe. He would send it in the morning, and that would be that. If they wanted to make something of it at Langley, then let them. The lights in the office were dimmed amber, and the tick of the safe's dial and the click of the tumblers reached out like a cricket in a dusky wood.

"There's a chill, Mr. Harper," said the Marine guard at the door, handing him the clipboard where he logged in the precise time of his retreat. "Like to freeze my ass off coming over."

"Have you seen autumn here?"

"No, sir. It don't ever get cold in Okinawa." The boy's face was as fresh as the pressed tan of his blouse.

"The leaves here turn color and drop to the ground. Can you beat that for exotic?"

"Yeah," said the guard. "The women here have a vertical smile just like roundeyes. Somebody lied."

"It's never as different as you might want it to be."

Harper didn't have a coat, and at first the wind went through his suit jacket with a shiver. But he walked a brisk pace, and soon enough he was warm again. He pushed himself, counted steps. Anything to try to take his mind off his problem.

They wouldn't let a man forget. Or more precisely, they always reminded him that he had forgotten. There were so many gaps in his memory of the Bartlow debacle. He had always taken it for granted that he would be able to recall all the details exactly if it ever became necessary. At the time, the killings burned into his memory so deeply that he could not imagine ever being rid of the details. But now necessity clouded his recollection, and he could only give Langley what he had left.

A taxi slowed and the driver waved him toward it. Harper looked away and paced on up the street into the wind. He took the rearward route where everything was shut up for the night. The light above a store spread out on the sidewalk, an obstacle. He stepped into the gutter to avoid it. When he spotted an open doorway ahead, he crossed the street and hugged the shadows of the wall.

When he reached the apartment he dragged himself up the tiled stairs, past the toys the super's boys left out ready for the next day's play. Balls, tops, airplanes, and assorted cars and trucks rose up the edge of the steps in an orderly file. On the landing their shoes also toed the line: slouching gym shoes, loafers, the wife's flats, the father's dusty boots. Harper stopped and picked up a model car. He spun the wheels against the palm of his hand. Some internal mechanism gave out an unconvincing whine. It was a Cadillac, the old kind with jutting fins and a grill like the jaws of a shark. In full scale it would have been much too large for Tokyo's cramped streets; you never saw them there. But, as if by some stone law, this unfamiliarity was precisely what attracted the children to the toy.

He put it back neatly in its place and mounted the last flight of steps. Through the door he heard the television going.

"Hello," he said tentatively. "It's me." The door was, as always, unlocked.

"Just a minute," she said from another room. He pulled off his jacket and laid it on a chair. It was warm inside, a fever breaking. He pulled out a handkerchief and wiped the moisture from his face.

She had the television tuned to a samurai tale. He stood across from it, leaning on the back of a chair. Low, angry voices challenged him. He did not understand a word. Their language could be so lyrical, but sometimes it was menacing. Haiku and battle cries, perhaps the language of every mighty civilization was cursed that way, sundered by all it tried to recall.

"What are you watching?" he asked as she entered the room and gave him a chaste little kiss.

"The only thing in English was something religious. This looked kind of interesting."

"Can you follow the story?"

"Not really. They're all the same, aren't they, like Westerns? I didn't really pay much attention."

Janet was fresh from the bath, her hair black with dampness, smelling of sweet lotions and balms. She wore the blue silk robe he had bought her as an apology for the weeks at the height of Birch's training when he had to be away. As she curled up on the floor beside his chair, the robe fell open slightly at the top, revealing the bright, inner curve of her breast.

She saw him looking down at her and closed the robe, but not too quickly. He did not want to mislead her about what he needed or could give tonight. He turned his eyes back to the set.

"Is this the one they always show at the film festivals?" he said.

"I don't think so, Richard."

"The one where it could have been a seduction, could have been a rape. I can't remember how it came out."

"I always believed the woman," she said.

"Of course. You would, should. But what if she wasn't sure?"

"A woman ought to know."

"I suppose we all ought to," he said, purposely putting weight on it.

"Did something go wrong today?"

"Everybody else says I have a poker face."

"Can you tell me about it?" she asked.

"As a matter of fact, I can. You know all about it anyway."

"Should we start with a drink?"

"I think so."

She poured his without asking his pleasure. She gave him what he needed, no more.

"What do I want?" she asked.

"Better have scotch," he said.

"That bad?"

"Very nearly."

Janet brought the tumblers, handed him his neatly on a coaster, and curled up on the floor. With his free hand he rubbed the long muscles of her neck, high up under the damp fall of hair.

"The station chief presented me with a cable this afternoon," he said. "I had to answer it ASAP. They've opened another inquiry into the Tri affair."

"I thought Bartlow was gone."

"He is," said Harper. "Now they're wondering about me."

"Oh, Richard."

It was their burden to have begun together in the worst of times. The memories of it were all mixed. Neither of them could remember the pleasure of those early days without also remembering the pain. As they lived together through his rage and depression, it was as if a strong field of gravity had taken over between them. They circled one another, each perturbation of one moving the other, close together but never quite overcoming the distance, like a pair of wheeling stars.

"The chief was really very kind about it, as always," he said. "He did not know who had pulled the file for review, but he said that it was inevitable."

"It's that stage of the war," said Janet. "We've lost it."

"The memorandum wasn't specific," said Harper. "Tell us what you remember. That's about all." The station chief had handed him the file jacket, coded in bright, warning stripes and stamped all over HUMINT—EYES ONLY. HUMINT, human intelligence, the red stripes were for blood. He had scanned the lines, his eyes taking in single words only—"murder" and "warning" and "failure." When he read the cable more carefully, there was not much more.

"I'm sure they're going over all the episodes," she said, "not only yours."

"The chief thinks otherwise," said Harper. "He thinks it's connected with the operation I'm running now."

"But there isn't any connection, is there?" she said, and Harper saw the concern in her eyes deepen, turn inward.

"Only this," he said. "If this operation goes rancid, Langley will have its doubts about me on the record."

The samurai barked and clashed, and Harper sipped from the tumbler of whiskey. Janet rested her head on his knees.

"You must get tired of my troubles," he said.

"No more than you do," she said. "I wish they would leave you alone."

"It's been our whole life, hasn't it?"

"We could have another, I suppose."

"I'd change," he said.

"Would you need me less?"

"Differently," he said. "Would that matter very much to you?"

"I've always known what I can expect," she said. "From the beginning I've known."

"Is that what we're resting on, that you've known the worst I can offer?"

"You know where the danger is," she said. "Maybe I hope you'll be able to hold me safe from harm."

"Young Lochinvar," he said.

"We're both professionals," she said. "We're like one another in so many ways. In a good partnership, everybody minds his business unless somebody gets in trouble."

"The distance is important to you, isn't it?" he said.

"I think it is, for people like us. We don't collide."

"That's a danger I depend on you to avoid."

"Don't be like that, Richard," she said. "We're comfortable. We fit together neatly. I'm not very likely to risk everything, and neither are you. It's a characteristic that runs in families with a lot to lose."

He touched her hair with the tips of his fingers, smoothed it out. Love and distance. The love could save you. But the distance was important too. It was a protection against betrayal and a recognition that they were always, though only in one quite narrow sense of the word, duplicitous.

"They were terrible questions that Langley asked," he said, because he knew this danger was something they could always share.

"No worse than you have put to yourself."

"Much worse," he said. "I've learned not to expect answers."

"What did they ask about Tri?"

"About Tri, very little," he said. "The dead aren't very interesting. They're beyond compromise or punishment. I think somebody got the idea that I hadn't raised a sufficient warning."

"Did they want you to blow sirens? In Saigon? Nobody paid any attention to them."

"They wanted to know what I told Bartlow and when. Did I let anything slip to the GVN? Why did I fly back to Saigon that night without telephoning Tri first?"

"Don't they know you got a medal?"

"It makes a good target," he said. "You know, you try to look back, try to remember. But it is so hazy. All the things you might have done, wanted to have done. The words you may have heard or said. The precise inflections. Have you ever tried to remember a thing exactly? A conversation. Something you've seen. I mean, to remember it when something has come up to make every detail crucial? It isn't only that what you want to have been the case gets in the way. You begin to wonder what you really wanted. I didn't want to fail."

"Of course you didn't," she said.

"But how much did that color what I said and did? At the same time I had my doubts."

"You always do," she said. "You are so very careful."

"I even had trouble deciding to ask you to marry me," he said. He was smiling now.

"You were just waiting until all the facts were in, until you saw exactly where the danger was," she said, smiling too.

"You can never be sure," he said.

"You have the strangest way of being flattering sometimes."

"Did you know that they had you checked out?"

"I assumed."

"Somebody put me on the distribution list," he said, the mood coming over him again. "I read the damned report."

"What did it say?"

"It doesn't matter. What's important is that I read it, and I can't say why. It would be so much easier if you never had to remember."

"Sometimes I look back fondly," she said. "I remember the Raffles in Singapore. We were so exotic."

"Maybe I wanted you for all the wrong reasons, even then," he said.

"Maybe it's how I wanted to be wanted."

"I wonder if duplicity is in our genes, dividing, repeating. Truth and falsehood, curling strands. Don't parallels always come together?"

"It's an optical illusion, Richard."

"It's a whole geometry."

17

The inquiry went nowhere. Harper's cable closed off the file. I only let the investigation begin in the first place because to have closed it summarily would have muddied up the record. The young man who initiated it has been transferred to a place that will temper his taste for intrigue. He understood nothing about what makes a man vulnerable. Harper was not one to be enchanted by Vietnamese of either stripe, theirs or ours. Two things must come together to occasion treachery—a weakness and a strength. Anyone should have seen that the latter element was in Saigon simply not present.

It was easy to close off the file because of the progress Harper was showing in setting up Kerzhentseff for the final stroke. Success is a great allayer of doubt, and the reports from Tokyo were exhilarating. It looked as though Zapadnya was about to step into the jaws of his own trap.

Under Harper's guidance, Birch had learned to play to the Russian's expectation of fear. It was so much more persuasive to Kerzhentseff than innocence. Birch put himself more and more in the Russian's control, all the while protesting the risks he was running. He became scared and submissive in equal measure as he made his regular reports on the progress of the new unit at Camp Zama. Harper even sent him to one meeting a little drunk. This earned Birch a reprimand from the Russian and a short lecture on the importance of keeping control. Birch, of course, was contrite. He said it was fear of discovery that made him nervous. And Kerzhentseff, to our infinite amusement at the time (and discomfort later), reassured him that the Americans were blunderers who were not so hard to fool.

The script called for providing a variety of details to Kerzhentseff: documents, observations, lists of names memorized from a roster. Much of this was accurate. Some confirmed what we believed the Soviets had learned already. Some was new and valuable to them. Counterespionage is the most potent exchange program ever initiated between adversaries. But nestled among the bits of accurate data were small hints about the Black Body detachment's mission. Kerzhentseff's interest was clearly piqued; from the Russian's questions, Harper could report that he was already edging toward the inference we wanted him to make.

The conventional method of driving him over the edge would have been to withhold the ultimate piece of information, to make him believe that he had discovered the secret truth by his own powers of deduction. Even the Mocking God does not reveal his tricks so blatantly. He lays out the clues cleverly—the strange reality of dreams, the tricks of fever, the unreliability of the senses. The surest deceit is in self-deception.

But, of course, Kerzhentseff was no conventional opponent, so the screw had to be given an extra turn or two to make sure it would hold. As Tokyo autumn turned to winter it was still just a bit too early to push the thing to its conclusion. We began orchestrating the independent bits. A surveillance of one of their deep-cover illegals showed itself to them in New York. A doubled asset in Vienna passed a vague and flawed warning. In the intelligence service of one of our European allies, which we had reason to believe was thoroughly infiltrated, the word went out that the Americans had scored a remarkable communications intelligence breakthrough but weren't sharing the take, the bastards. Details were left to the fecundity of rumor.

Meantime Harper busied himself by collating what the operation had revealed about the Soviets' methods. He was impressed by their deftness. If Birch had not been under our control, Harper had no question but that the meetings would have gone undetected. Our dead drops were crude in comparison with their multileveled cutouts. There was a delicacy in their method that was more characteristic of the nation that produced the Bolshoi than the one whose successive five-year plans only yield longer queues.

He had more than an academic interest in his side project. He mapped out a regular Michelin guide to their safe houses and meeting places. He sent out spotters to try to determine the location of future rendezvous. Kerzhentseff seemed to favor the warehouse district where the first interrogation had taken place, and on one occasion the spotters actually detected the Soviets moving their equipment into a building the morning before a meet. Harper deeply wanted to have one of the meetings wired for sound. The official reason was to run stress on Birch on a real mission and see how well the training worked. The private reason was that he wanted to hear Kerzhentseff's voice.

It was this obsession that brought him into close consultation with the Agency's chief ears, Ralph Braybeck. And it was from him that Harper learned, quite by chance, about the fire.

Braybeck had taken upon himself the job of planting the microphones in the brothel before Birch's last contact with the man who called himself Nowicki. His only regret was that once he got the devices out again, operational caution prevented him from returning to share the favors of Kokura. This did not keep him out of the neighborhood entirely, though. There were many other establishments with similar attractions, and one night during one of his randy walks he passed by the building.

"Gutted," he told Harper idly over lunch. "Burned down. Nothing much left of it really. Four walls. Even the roof's gone. Hell of a mess."

"When did this happen?" Harper asked.

Braybeck held up a finger as he gulped his Coke to slush his mouth clear. Then he wiped his lips clean on his shirtsleeve.

"Found some interesting places in Shibuya, near our safe house," he said. "Out of the way. Strictly locals. No GIs spreading the drip. Lusty broads."

"When?" Harper demanded.

"Haven't been there for maybe ten days, two weeks, like I said. Out exploring virgin turf. Could have been anytime in there."

"Did you hear anything?"

"Didn't push it," Braybeck said. "A little close to the bone, don't you think? Asking questions, that's your side of it. Me, I just listen."

"No talk at all on the street?"

"Talk," he said. "Talk. Sure, there's talk. They say some of the poor girls died. Nothing specific. Big tragedy is all. Damned waste. Kokura was four stars."

Harper left him and made his way to the tiny office where the Scoutmaster directed his teams of spotters. He was a man who chose his words carefully because he used them so rarely. As a rule, the Scoutmaster's mission was to be neither seen nor heard.

"Take it easy, my friend," he said in an agony of slowness. "The uh . . . We'll just have a peek at the uh . . . the logs here."

He pulled out a loose-leaf notebook coded in bright, forbidding colors like something radioactive.

"The uh . . . How far back did you say?"

"Ten days. Two weeks," said Harper.

The notebook pages snicked and settled under the Scoutmaster's left hand. Harper leaned over for a look, but the Scoutmaster pulled the notebook into his lap to discourage him.

"The uh . . . need to know. Other ops in here. The uh . . . You understand."

"Sure. Go ahead."

The Scoutmaster studied a page, and just as Harper decided he must have found something there, he slowly turned over the leaf. Nothing had been reported to Harper, but then again it was not unusual for Birch to wander the city. The spotters might have seen no significance in it. If there had been no contact, they might have paid no attention. But Harper was not interested in whom Birch might have met; he was interested in what Birch might have seen.

"The uh . . . The uh . . . He's stayed pretty close to home," said the Scoutmaster. "Hasn't been anywhere near the brothel."

"Thanks," said Harper, but he was far from finished with the matter. His anxiety had slaked, but he put no faith in accident. Where there's fire, there's smoke. And Harper was determined to know what it was intended to obscure. His first step was to visit the Agency's contact in the local police.

"You will have tea?" he asked when Harper arrived.

"Certainly."

He was a youngish fellow, educated in the United States as an undergraduate and then later under the auspices of the Sisters at Quantico. Mr. Mikima was his name, Mickey to the Sisters. He was square of head, like a blocked-out sketch that has been abandoned incomplete, and in build compact rather than small. He spoke excellent English, lightly inflected but in no way strained. Harper judged that he would have been born during the war. He would have grown up under MacArthur's benign dictatorship and known the Americans as natives in his land. This made the unsovereign relationship easier.

"Please make yourself comfortable," said Mr. Mikima, directing Harper to one of two couches in the short leg of the L-shaped office. They crowded so closely to the table between them that Harper had to move crabwise to get to his place, like a latecomer at the theater.

"I'm glad we have this chance to meet, Mr. Harper. I have known many of your colleagues, but somehow our paths have never crossed."

"I guess I've managed until now to stay out of trouble."

"You are a diplomat," said Mr. Mikima in a way that made it seem a compliment, not just a recognition of Harper's cover and immunity.

A secretary brought the tea already poured into two fragile cups. Harper sipped; as always, it was too sweet.

"The dead have all been identified," said Mr. Mikima. "Perhaps you would like cream."

"This is fine," said Harper, taking more of the spicy syrup.

"We were fortunate that they had all been known to us before."

"Arrested?"

"From time to time," said Mr. Mikima. "Like you, we keep up appearances and check for disease."

He lighted a pungent cigarette and then handed Harper a typed list in English. The names of the victims: Kokura, Toshi, the old proprietress, and four others. He opened a file on the table and slid the papers from a neat stack. It was drab, yellowish newsprint.

"Those records can't hold up for very long on that kind of paper," said Harper.

"They turn to dust. It is a way of ridding ourselves of the past. I know that must amuse you."

"We never throw anything away. We think this is a duty to posterity."

"You are a very young nation," said Mr. Mikima.

He riffled through the pages and found the one he wanted. Harper waited patiently as he read it to himself, watching the man's eyes scan with the indifference of history the ugly details.

"It was clearly arson. Traces of kerosene and flash powder. A quick, effective job of it."

"Did anyone survive?"

"The fire had apparently spread throughout the building by the time the first trucks arrived. They were very lucky to have contained it to the one building. No one survived but those who set it."

"A bad one."

"Worse," said Mr. Mikima. "There is evidence that the bodies had been cut before they were burned."

He handed Harper photographs of the grotesque remains.

Harper forced himself to look at each of them, then he put them aside.

"It is the one part of the record that needs no translation," said Mr. Mikima.

"In the States, I believe, these fires are usually connected with insurance."

Mr. Mikima took the photos back and studied each before returning it to the file jacket. Harper caught him wincing, but it was perhaps only the smoke from his cigarette.

"I don't think your organization pursues insurance fraud," said Mr. Mikima.

"You are a diplomat too," said Harper.

"Yes. Thank you. If you would like a copy of the file I will have the girl make one. I'd appreciate your holding it closely. Basically it shows that from where the bodies were found they were probably killed in different rooms. This we have had to infer, since the roof and floor collapsed. The stab wounds, however, these are conclusive."

"Suspects?"

Mr. Mikima did not answer at first. A serious look darkened his face, and then it gave way to a strange smile. Smiles were, in these matters, acts of state.

"It would have been more than one person," Mr. Mikima finally said. "They were very well trained. A team. I would say they had quite a bit of experience and a well-conceived plan. Professionals."

"So it could have been purely business," said Harper.

"Our criminal enterprises, if you will excuse me for saying so, do not act as yours do. This is not their fashion."

"Not violent?"

"They would prefer a single disfigurement. Perhaps the prettiest girl. An ugly scar across the sweet face. A living proof and a warning, if this was the purpose."

"Passion then?"

"We are no different than you, Mr. Harper. We are very methodical, but our passions do not have plans."

"Yes. I suppose not."

"Perhaps you will allow me a question."

"My guesses?" said Harper.

"Or your conclusions, if it has gone that far."

"I'm afraid I cannot be as accommodating as you have been."

"Reasons of security."

"I can assure you that no one in our control had any part in this."

"This," said Mr. Mikima, a little too quickly, "goes without saying."

"I prefer to make it explicit," said Harper. "I am not asking you to terminate your investigation."

"Then you assume it was one of your adversaries."

"I would prefer if you drew no conclusions from my interest."

"It would be quite fruitless to speculate, Mr. Harper. You have so many adversaries."

Mr. Mikima dropped his cigarette into the ashtray where it sent up a straight, unwavering signal in the silent air between them. Then he closed the file folder, and the breeze shattered the thin plume. He slapped his hand conclusively on the manila jacket and stood up abruptly.

"So," he said.

Harper wanted evidence that Kerzhentseff had done it. Witnesses. Maybe some girl who had worked there once but didn't any longer. Maybe she saw them come in when they cased the place, was frightened, fled. There had to be traces. A street boy's tale, who had nothing to gain or lose. The regulars at the pachinko parlor across the way. But Harper could not investigate. It would have been just like Kerzhentseff to have killed the women just to see who came to poke around in the remains. Zapadnya was quite capable of baiting the trap with human flesh.

The stink of death was on the Black Body operation. But no matter how much he gagged on it, Harper knew better than to make the same mistake he had made after the murder of Tri.

He had to treat the thing coldly, professionally, to live with the uncertainty and the sinking sense of responsibility that went with it. And, to his surprise, this time it came much more easily.

He did what he could, reviewing the file and the tapes they had made at the brothel. Without success. In the end he prepared a short report of what little he had learned and sent it to the station chief. The evidence was scant and jagged; it fit together in any number of ways.

"We must assume," said the station chief, "that the fire was their doing. And that means they had something to achieve. The Soviets are never purely gratuitous."

"I'd say they were afraid of what the Slav may have let slip in pillow talk," said Harper. "But what would the Slav have known? Not enough to kill for. Unless they thought the girls were ours."

"If they thought that, they would never have done it."

"Only innocents die," said Harper.

"Death confers innocence," said the chief. He gazed up at the bold calligraphy on the opposite wall. "Perhaps they meant it as a signal."

"To us?"

"Let's run the possibility out for a minute."

But Harper knew where it would lead. All inferences ended in the wilderness.

"If it was a message, they must have thought we were there to receive it, that we have been watching the brothel," Harper said.

"Because the Slav frequented it."

"Then they suspect we have discovered Birch."

"Yes."

"But what would the message be trying to say?" said Harper.

"That is the puzzle," said the chief. "Perhaps a question: 'Are you there?' "

"I've thought about that. If we are there, then we would be expected to withdraw Birch for his own protection—if he is under our control."

"We would see them closing in and have to respond," said the chief. "This would be their way of being sure there is no trick."

"Then there is nothing for us to do but to leave Birch in the game, keep the fact of the fire from him, and charge on," said Harper. "The fire didn't even come to our attention. Why should it have? A police matter. Nothing more. Because we don't have a line on Birch at all, let alone the Slav. He was small beer, not worth our surveillance."

"Nobody here but us chickens," said the chief. "No matter how it plays, Birch must not be told."

"I think this is the best explanation," said Harper. "I'm satisfied."

"You know, I've always preferred the Asians," said the chief. "I realize you have had your go around with them, but that was a bad class of Oriental—we'd already corrupted them. Generally, the Asian does not go in for such blunt strokes. He likes to use his opponents' wits against them. He acts against the synapse, not the sinew. It's the difference between chess and a game of Go. He does not send out pieces to knock you off the board. He surrounds you, cuts you off, leaves you stranded alone with your doubts."

"I'm not sure which is worse," said Harper.

"The Asian way, without question."

"You said you preferred it."

"More challenging," said the chief. "Have you noticed that Kerzhentseff has a Mongol look? Just a trace. Many of their best agents seem to."

"Subtlety and violence combined," said Harper.

"Let's see how it plays. What if his real purpose were to reassure us?"

"He didn't exactly succeed."

"Didn't he?" said the chief. "Remember what we aren't questioning here. Suppose he has turned Birch but wants to cover it. Suppose Birch has told him we know all about the Slav and the brothel. He knows we have put together something quite large and expensive to support whatever deception we intend to run against him. He can count, after all, and the sheer number of personnel we put at Camp Zama tells him something about the importance we place on this operation. He can tell nearly as much from our lies as from the truth, so long as he knows which is which. He's curious. He wants to keep Birch on the case, to see where we are driving. So he has to make it clear he does not take Birch for granted."

"I think we're safe with Birch," said Harper.

"You must, of course," said the chief. "No matter. Let's continue the exercise. He wants us to believe that he does not know Birch is in our hire nor even for sure whether we are on to him or not. But he is fully aware of his own reputation. He must make the effort not to be duped. So he sets the obvious trap."

"Self-parody," said Harper. "It seems a little remote."

"That, of course, would be its virtue. He kills the girls and sets a fire so that we will assume he is testing Birch's bona fides. But we do not fall for it. He is not one to underestimate us, you know, whatever he might be telling Birch about our competence. We see that he is trying to draw us out, and we pat ourselves on the back for refusing to bite the lure. Meantime, Birch is spilling his guts to Kerzhentseff, and we don't even suspect it."

"I guess we do now," said Harper.

"In a sense," said the chief. "One learns to suspect everything. But the point is to choose what fear to act upon."

"We could lean on Birch," said Harper.

"Flutter him on the machines we've taught him to beat?"

"That's a problem," said Harper.

"We don't want to drive him over to the other side if we believe—and I think you do—that he isn't there already. And if we were to scrap the operation, well maybe that is what Kerzhentseff is looking for too. I don't need to play out that string, do I?"

"A mirror reflecting a mirror. It ends in a vanishing point."

"That's the common image," said the chief, his splayed hands coming together before him like vectors of force. "It all comes to the same thing in the end. There is no advantage in giving Kerzhentseff an answer to his signal. The best we can do is keep an ambiguous silence."

"And wonder."

"And wonder," agreed the chief.

"You have reviewed Birch's file," said Harper. "Do you see any vulnerability there?"

"All sorts of things. He is very young and naïve. I've seen good assets, solid gold in the vault, who have decided that to save mankind they must play the middle, the vanishing point, as you put it. Idealism unalloyed with loyalty.

"And Birch has his reasons to be bitter with us. We forced him into the brothel, after all. Before we came along, he led a rather pleasant, uncomplicated life."

"I think it excites him."

"Another weakness," said the chief. "Or perhaps a strength."

"We have no choice but to trust him."

"Exactly right," said the chief. "There is something in everyone's file, you know. I've read yours."

"What did you find?"

"An optimist," said the chief. "Very dangerous."

"I thought the suspicion was behind me. I thought the inquiry was closed."

"It is. It is," said the chief. "I was only trying to make a point. I know what's in my file too. How it could be read. And I could make the case against myself quite easily."

"Do you?"

"All the time, Harper. All the time."

"At least I'm not alone."

"We all are—together," said the chief, and it was difficult for Harper not to take the expression on the man's face as fraternal.

18

When Birch's voice came on the telephone and spoke the single, distressed word, it was all Harper could do to keep himself from trying to talk Birch out of it. But the arrangement was absolute. The code word, in these circumstances, meant move to the prearranged meeting place—and move now.

Janet came half-awake and stretched.

"Problem?" she said.

"Might be," he said as he felt his way to the chair where he had hung his suit. "Damn," he said, his toe encountering the hard leg of the bed.

"Time is it?"

"Late. Past two."

"Unh," she said and slid back to sleep.

It wasn't the hour that disturbed him; it was the date. It was simply too late in the play for a crisis. He gathered his clothes and took them to the bathroom where he could turn on a light. When he was finished dressing, he fumbled with the knot of his tie, avoiding his own eyes in the mirror. Then he splashed cold water on his face, which only articulated the fatigue, and went to the kitchen to dial the Scoutmaster's office on his secure line.

"He's out," said the spotter on duty. "Wandered through the door a few minutes ago. Funny time of the day for a walk."

Harper gave the man the location of the park and warned him to keep his men well back.

"Lose him if you have to, but don't let them be seen by him or anybody who may be following him."

"I was going to give you a call. Hated to trouble you, though," said the spotter, "especially if it turned out he was just going for a pack of cigarettes."

"He doesn't smoke," said Harper.

"Sorry."

There had been a contact earlier in the evening, the one they had been waiting for. Birch had received his signal from Kerzhentseff and gone to an outdoor market in Yokohama, as arranged. He had stood at the orange bin, testing the fruit and making way for the Japanese ladies who were there to buy. The messenger came, made the pass. The only thing odd was that he had lingered. There was a little conversation, very unorthodox.

"Any of your people know what was said?" asked Harper.

"The soundmen had directional mikes, clever dogs," said the spotter. "Ready for anything. But the damned things didn't work. Scrimp and save. Make the stuff last, they tell you. Push the button in a pinch, though, and nothing happens. Makes you wonder about the missiles, doesn't it?"

"What was our man's reaction?"

"Nothing special," said the spotter. "Says here he bought some oranges. He moved off smartly after the meet. Straight home to momma."

"He's going to be evasive tonight," said Harper. "What I want to be sure of is that the others aren't behind him."

"They're on him?"

"They may be. I'll be meeting him in a park. If your people make me there, be sure they stay away. Unless, of course, they see the other guys on his tail. Then I want to know it."

"Roger that. You want any backup?"

"No. Definitely not."

"Notify anyone? The chief?"

"He'll find out soon enough. Leave him sleep," said Harper.

"Good hunting."

"Weasels hunt at night."

"And owls and Americans," said the spotter.

Harper made one last check in the bedroom. He found his keys in the tray on the bureau and whispered good-bye to Janet. She had a word, but he did not understand it. It belonged in some other nightmare. He left quickly and went for his car.

The chief had devised the endgame strategy, and though he was Asian only by experience, he played a fair game of Go himself. The plan violated all the common assumptions, and that was its elegance. It was too direct to be seen as contrived. In a world of ambiguity, the obvious could be the cleverest sleight of hand. They had been preparing for it for weeks. Birch had to be given a formal transfer to the dummy detachment. If this seemed to Kerzhentseff, as it must have, too good to be true, the chief had a further move that would make it seem too good not to be. A rather complete personnel file was created to make the transfer orderly. Birch even received an Army Commendation Medal for his skillful liaison work. The medal was presented in Colonel Robertson's office, and a photograph of a beaming Birch and his proud wife appeared in the next issue of the camp newspaper.

The chief had been delighted by this detail. When Colonel Robertson sent a copy over, the chief read it and slapped his hands merrily on the desk.

"Perfect," he said. "Just perfect. I want you to savor it. I've noticed you looking at my bookshelf from time to time, the fiction and drama there. You probably thought of it as a pretension. But the writers, they have all been there first. Illusion, Harper, illusion. The tiny details that seem gratuitous or irrelevant but then finally reveal their meaning. I think it was Nabokov—a Russian, you know—who noticed that when a butterfly has to look like a leaf its markings are appropriate down to the tiny grub holes that no predator has the visual resolution to notice. It isn't survival, Harper; it's exuberance and art."

"Perhaps you both underestimate the predators," said Harper.

"I prefer to think of it as creating an audience, creating it by doing the thing with infinite care and imagination." He held up the newspaper at arm's length and regarded it. "Beautiful, don't you think?"

Beautiful, perhaps, the way a grub hole is. Beautiful if it has the proper results. For Harper the thing was at best not an art but rather an evil redeemed by consequences. And that made failure the greatest sin.

This haunted Harper as he responded to Birch's telephone call. They were too close to success for Birch to have second thoughts. Kerzhentseff was sniffing up the trail of inferences that had been so carefully laid down for him, and it was imperative that he not be diverted. All the evidence pointed to the conclusion that the Soviet had already connected the dummy detachment with the system we called Black Body. His questions to Birch were aimed at determining whether the unit was having any success, and they were getting more and more insistent. In the meantime we had word from Vienna that our people had stumbled upon a new deep-cover agent in a conventional tryst with his Soviet handler. If he had been fitted out with Black Body, such a meeting would not have been necessary. Similar but less conclusive sightings had been made in Bonn and New York. It was beginning to appear the Kerzhentseff had taken a half step in the right direction: the systems in place were allowed to continue functioning, but the expansion of Black Body had been put on hold. We gave him a little more to worry about by leaking the code name in Beirut and by running a giveaway operation in Washington followed by a damage assessment as if we had picked the compromise out of the air.

The streets were nearly empty as Harper got under way. A few night delivery trucks were parked here and there. The men hurried about their work, conspirators afraid of the dawn. Harper, eyes mirrored rearward for signs that he was being followed, drove a circuitous route. A taxi stayed with him for a time and then finally peeled away. Just to be sure, he turned into a quiet, residential section where everyone safe was safely inside and the only signs of attention were the lonely, lighted windows of insomniacs. No headlights followed him.

Up ahead, the park darkened part of the side of a hill patterned with streetlights, a cloud obscuring stars. He parked the car well away from it and went the rest of the distance on foot. The buildings here all had the look of the horizontal—long, bold lines to divert the eye from the heights, as if to give the illusion that in this crowded place there were great spaces to squander. The air was chill and damp on his face, and he trimmed from one light to another to stay in the shadows.

The park itself was silent, too cold now for the insects to be singing, too dark for birds. But he knew that others were there watching, his people and maybe theirs as well. He shivered as he found the footpath and followed it into the trees. The path verged on a lagoon, and on the opposite bank a single lamp shone out across the water in perfect duplicate. He felt that he was seen by eyes unseen. He crossed an arching wooden bridge, his footsteps reaching out to announce him. There was nothing to do but to hope that the eyes were protective.

The spot they had chosen was a cul-de-sac off a broad meadow. There were a few benches there for taking the sun in the daytime and for lovers in the night. He approached them softly from the tree line like a sniper, but they were empty, glowing with dew in the moonlight. Harper put his back to the thick trunk of a tree and rested there in ambush.

But he did not feel alone. Every rustle of a small animal or the breeze in the high branches was to him a stalking man: theirs or ours or something in between. It was hard enough to distinguish them in the bright glare of the day. But now at night in the forest he was both hunter and prey, betrayer and betrayed. The woods were alive with duplicity, and imitation was the highest form of scorn.

Something moved behind him. He turned. There was nothing but the darkness. Or had it been a human seeking him out, stalking from tree to tree, a knife ready in his hand? No, he told himself, drawing a long, steadying breath. That was not the kind of violence it was reasonable here to fear. Only the innocent die, and in the enchanted woods everyone is a party to the crime.

Clouds rolled across the half-lidded moon over the clearing, a thick living tissue, then parted again. Suddenly, into the dazy glade, Birch emerged from the trees. Harper forced himself to look beyond the man into the shadows. No movement. He seemed to have come alone. Harper backed slowly into the woods.

The meeting place was under the bridge, two trolls in their habitat. Harper reached it first. He felt his way among the refuse and bottles that littered the bank. His hand slid along the wet, mossy brick wall. From above he heard Birch slide clumsily down the embankment.

"Over here," whispered Harper. "What's this all about?"

"We have to talk," said Birch, close enough now that Harper could smell the problem.

"You've been drinking."

"Never mind me," said Birch.

"I've got to mind you. You are my business. And it looks like you are determined to make a mess."

"Shut up," Birch shouted for all the forest prowlers to hear.

"Softly," Harper whispered.

"Can't you just listen for once? Can't you?"

Weakness always showed itself too late. That was what distinguished it from strength. Harper touched Birch's arm in the darkness, held it steady.

"You called and I came," said Harper, "the way I promised I would whenever you said the word."

Birch freed his arm and backed away under the high, arching timbers of the bridge.

"You promised," he said.

"I don't understand what is wrong."

"What have you done to them?" Birch demanded to the last edge of his breath.

"Done to what?"

"Not what," said Birch. "Don't you see? Not what. Who. The girls. What have you done to them?"

"We've done nothing."

"Tonight, at the market, the man said I didn't need to worry about the whores anymore because my friends have taken care of it. He said Kerzhentseff thought I'd want to know."

"I'm afraid he was talking about your Soviet friends."

"Then it's true."

"Yes it is. They did it by fire."

"For what?"

"To keep us from finding out about your contact with No-wicki, I imagine," said Harper. He could have gone the whole distance. Chess and Go, laying out the alternatives with the last one now Kerzhentseff's trick to send Birch running out in the middle of the night so that his handler would hold his hand. But Harper did not. If Birch's hand needed holding, he would hold it.

"My God," said Birch, slumping against the brick wall. "They did it to protect me from you."

"That's what we assume," said Harper. He chose one way of looking at it, and it did not matter which he chose, since all of them led to pain.

"It's my fault," said Birch.

"I don't think they needed your encouragement," said Harper. "I think it is second nature to them."

"I was so smart," said Birch. "I let Kerzhentseff think I was scared. I let him think I wanted nothing more than to be free of you."

"It was a good strategy, Birch. It worked well."

"We killed them."

"The girls," said Harper. "That was unforeseen."

"I would have been better off going to Nam," said Birch.

"You are at war here too, Birch."

"I'm a coward."

The little stream bubbled over a protruding rock, and Birch wept softly. It disgusted Harper. But the Agency wants only cowards. It cannot trust the brave.

"It takes nerve to meet the enemy man to man," said Harper.

"With lies."

"With whatever comes to hand."

"And whoever they hurt."

"What do you want me to tell you, Birch? That it's all OK? That it's worth it? I don't know how you measure the national interest against a single soul. But I do know that the nation always has its way."

The weeping had stopped. Birch sniffed unmanfully, taking short, uneven breaths. All they ever want to hear is that it could not be helped, that it was, in one sense or another, necessary: chance or fate or nature. And someone is always ready to say it because someday everyone needs that solace. Excuse is reciprocal, just like any other falsehood.

"We are different from them, aren't we?" said Birch.

"We do what we must," said Harper. "Nothing more."

He wanted to seize Birch by the shoulders and shake him and tell him the simple, ugly fact: that we are all brothers and that what is done is done in our common name. But this was not the time. Truth is for when there is nothing left to lose.

"When are you to meet him again?" Harper asked.

"Tomorrow night."

"I want you to go home and get some rest then. We have a lot of work to do before you go."

"But the dead girls," Birch sobbed.

"We can't allow them to have died in vain," said Harper, the last, inevitable exculpation. It soured on his lips. You can always trust the dead to serve. In silence, their loyalty is unswerving.

"I would kill him for what he did," said Birch.

"Of course," said Harper. "But we can't allow that. There are better ways of retribution."

"But he will never know," said Birch.

"What could be a better punishment? Now I want you to leave ahead of me. I will wait until you clear."

"One thing first," said Birch, coming very close, his whiskey breath heavy in the cold air. "We would never have killed them, would we? I mean, whatever else, but not that, no matter what the reason."

Only in retrospect, long, long after he watched Birch slouch off into the obscuring night did Harper begin to wonder whether he might have given the wrong reply. Only when he began to try to puzzle out what might have driven Birch over to the other side did he wonder whether he had failed to make the distinction clear. For his own part, Harper was disgusted by the very thought of the killings. He had seen the horror twice and never wanted to see it again. But he suspected that this was his weakness and that Birch had experienced enough already to know that any absolute was a lie. So when Birch asked for reassurance about the difference, the saving difference, Harper said it the only way he found plausible:

"I don't see how we could think it would be necessary."

The next morning the signs were not good. Birch, always punctual, was not at the detachment when Harper showed up tired and sharpish.

"No sweat," reported the spotter. "He hasn't budged. He's inside bitching at the old lady about the way she did his eggs. Slow start is all. I'd say he like maybe had a little too much to drink last night."

"I want him here," Harper snapped into the telephone. "There's no time for screwing around."

"You want, I'll go grab him."

"Of course not."

"Maybe he's not the only one feeling a little frail this morning, if you don't mind my saying."

When Birch finally did arrive, Harper knew that something was seriously wrong. It wasn't only that his usually knife-edged khakis were rumpled or his eyes webbed with veins. Gone was the open, eager look of a child who wants to learn and to please and who does not fear that the lesson may be cruel in equal measure as it is true. Here now was a man, not wizened and not quite worldly, but still a man who has been initiated into the steady-state vision of standoff where every gain is also a loss. Harper drew him aside into a corner where they could be alone. The technicians busied themselves with the crystalline certainties of their circuits and engines.

"I thought there might be something more," Harper said, "from last night."

"Like what?"

"You look as if you didn't sleep very well."

"That's my problem."

"I had a hard time too. If it weren't for the meeting, I'd have given us the day off," Harper said, rubbing at his neck.

"Duty first."

"Always," said Harper, with an unreciprocated smile.

"I don't give a damn about the meeting," said Birch.

"I can see how you might feel that way."

"You don't see anything," said Birch. "To you it is just words you say to get what you want. Words you say, words you don't. You knew about the fire and you didn't tell me."

"I didn't want to upset you."

"Because I might just say the hell with it and walk away. Don't try to tell me it was to protect me. That's what Kerzhentseff had in mind. It's a word that covers a lot of ground."

"The mission is very important. I won't say it doesn't matter to me," said Harper.

"Now you are going to tell me about war again," said Birch. "The bad Russians. The good soldiers. Well, maybe it is the war that's bad and not the people—on either side—who are fighting it. Maybe it's the idea that everything is more important than what you feel, than life. Maybe it is a sin to lie. Don't you ever wonder whether there are really any exceptions?"

"I'm not a religious man," said Harper. "Sin isn't what I wonder about."

"Only what is necessary," said Birch.

"That's right."

"And what you can get the hick from Cleanthe to do for you."

"That is up to you, Birch. It always has been. You know as well as I do what is at stake. I cannot make you take the risk any more than I can make you sleep at night. I think we'd better get started."

Birch hardly seemed to pay attention when Harper showed him the documents he was to turn over to Kerzhentseff. Harper read the key one to him aloud, hoping the spirit of the thing might take hold. It was quite vague about the technology of the interceptions—had to be, since there wasn't any. But it did describe, in appropriate bureaucratese, plans for establishing a number of similar Black Body receiving stations and praised the detachment's work as having cleared the way for quick approval of the large-scale program.

Harper explained the circumstances under which Birch was to say he had access to the materials. Sunday duty. No one else around except for a few men manning the devices in the back. A file discovered during security rounds. It was loose in the commanding officer's desk drawer, a commendation letter. He must have been reading it over to himself. Birch copied it, secreted the copies, locked away the originals and reported the incident dutifully.

"He's not going to believe that shit," said Birch.

"He will if you can convince him."

"You're setting me up." Birch sprang out of his chair. One of the technicians grabbed his arm to restrain him, but Harper gestured to set him free.

"You don't believe that," Harper said.

"I don't know what I believe and what I don't. Maybe your machines will tell us both."

"It's been a tremendous strain on you. I appreciate that. You feel very much alone right now. You don't know it, but we are all there with you. Still, it's a scary place. And nothing is familiar."

"There isn't anybody else," said Birch, suddenly tired and resigned as if he had exhausted his last reserves.

"I want to take you through the flutter drill again," Harper said, matter-of-factly, as if everything were settled and facts were the only matter.

"Whatever you say," said Birch.

The technicians strapped on the electrodes as Birch slouched in his chair, neither helping nor resisting. The straps and wires he suffered like instruments that have lost their power to torture.

Q. Is your name Jerry Birch?

A. Yes.

Q. Are you an enlisted man in the United States Army?

A. Yes.

"This is very strange," said the technician. "Take a look."

Harper took up the strip of paper and traced his finger along a jagged, erratic pattern. Birch had bridled at his own name.

"We'll have to switch to the backup machine," said Harper. "This one is on the blink. Bear with us, Birch. The story should be easy enough for you. Just relax a minute."

"I don't think there's anything wrong with the machine," said the technician. "I checked it out this morning."

"Change it," Harper ordered. He knew what Birch was up to and wanted to give him his little victory and time to reconsider it. He winked conspiratorially at Birch, who accepted it blankly.

Q. Is your name Jerry Birch?

A. Yes.

"Same thing," said the technician.

"Can't you get the damned thing to work?" Harper said. "Maybe there's a problem on the electrical line."

"We run off the generator."

"Then just fix the sons of bitches."

"It isn't the machine."

"Humor me," said Harper, ripping off the tape and studying the same troubled pattern there.

"Birch," he said, "if you are playing with me, I want it stopped."

"Now the accusations," said Birch.

"I'm disappointed that you feel you have to do this."

"You asked me who I was and I answered."

"We know who you are. And we know that you can control the pens."

"You tell me whether I was telling the truth then," said Birch. "Maybe I don't know anymore."

"Kerzhentseff will not be amused."

"I'm not here to keep him happy—or you."

"Yes you are, Birch. You're here to make him the happiest man on earth—the perfect fool."

"I thought that was my title."

"I'm going to take a little walk," said Harper. "You get yourself together and he'll check out the machines. When I come back we'll go through it again. And if there's a problem, I'm going to call the operation off. Just remember one thing, Birch. If we walk away from this thing now, if we lose our nerve, the fire will have saved Kerzhentseff from defeat. And in a twisted way, you see, it will have been justified."

The rain had stopped, and the air was still and cold. Harper stood just outside the door and waited. He was not bluffing. It was his decision, and he was ready to make it. He would explain himself in the language of operational risk—the danger of compromise, the considered judgment that ambiguity and incompletion were preferable to the possibility of a total failure. But there was more to it. Harper was quite candid about it in his personal notes. He was simply unwilling to let Birch risk his life.

Harper left the detachment compound and wandered through the camp—counting his steps, the cracks in the sidewalks, anything to tick off the time and keep himself from questioning the decision he had made. Soldiers hung out on the steps of the barracks, smoking, playing grabass, waiting to be told to die. They were so young. Only boys are willing to do the horrid things required of them—mean and believing little boys. But Birch had suddenly matured. They often did, under fire. Harper walked for a half hour before allowing himself to return.

"Is everything set up?" he asked, stripping off his coat and not giving Birch so much as a glance. It had come to something automatic, the toss of the dice, the calibrations of a machine.

"All ready," said the technician. "Always was."

"I'm ready too," said Birch, and Harper heard defiance in it. But that did not matter. The polygraph would tell. And in the coldest, most mechanical voice, he asked:

Q. Are you Jerry Birch?

A. Yes.

Q. Are you an enlisted man in the United States Army?

A. Yes.

Q. Is this a copy of one of the documents you found in your commanding officer's desk?

A. Yes.

Q. Do you believe this document is authentic?

A. Yes.

Harper pulled the lengthening piece of paper between his fingers and scanned the lines. The curve was as lazy as a summer hillside, as real as a mirage in the sun.

"It's OK, isn't it?" said Birch. "We're on, right?" Harper looked at him and saw that perhaps there was just enough of the child left in him to bring him through.

"Didn't fool me," Harper said, and rewarded his student with a laugh that was very nearly genuine.

19

I cannot explain the hunch that gave us ears at the meeting that night. Harper had been plotting the Soviets' habits since the first contact with Kerzhentseff, and the way he describes it to admirers, the pattern simply appeared to him in the end. I have looked at his records, trying to reconstruct his thought process in hopes it might be taught to others. But to no avail. His hunch was simply inspired, which is a characteristic of both genius and plagiarism.

Whatever afflatus it was that whispered to him, the soundmen made quick work of getting the device in place. And when the spotters reported that the Soviets were indeed moving in their equipment, Harper hurried to the observation post.

He was worried about Birch but not enough to call the thing off. The man had snapped out of it remarkably quickly. He had learned the script with his usual ease. And he had again demonstrated his mastery of the polygraph. There was no choice but to go ahead.

The observation post was a grim series of offices in the rear of a vacant, two-story factory. From it, they had a view of the warehouse where the Soviets were set up and the alley behind it. But more important, they had a line of sight to the listening device, which was secreted in a window frame, and they could switch it on and off with an electronic pulse to evade any Soviet sweeps. The factory smelled of creosote and damp rags, and the walls were scrawled with calligraphic graffiti and drawings universal in their depravity.

Braybeck met him downstairs and showed him to the rooms where he was supervising the final adjustments of antenna and gear.

"Some setup, huh?" said Braybeck. "In there we got to keep it dark by the windows. This place is officially empty. But you can slip in for a peek if you want."

"Any sign of our man yet?"

"Nothing. The bad guys are hunkered down across the street. We think they may already have swept."

"Any chance they found our bug?"

"Only way we'll know is when we turn it on. But I'll tell you this, I couldn't find the little bugger if I didn't know where it was. And I'm the best there is."

"You're going to have to learn to overcome that diffidence, Ralph," said Harper.

"I'm just saying."

"Where's the lookout?"

"Through here. Just be sure to close the door behind you. We're using the middle room as a light lock. The connecting door will be straight ahead in the dark."

Harper glanced around at the activity in the brightly lighted, windowless cell. Men with earphones silently fiddled with the dials of their boxes. Others prepared a sequence of tape machines.

"You gonna be all right?" asked Braybeck.

"What?"

"You don't look so good. Look, we got this thing knocked. I mean, it is no sweat. Easy as switching on the old FM."

"I'm not worried about the reception," said Harper, and he entered the darkened anteroom. When he opened the second door he could see the spotter, hunched beside the window, turn and raise his hand in a greeting. Harper took the other side and peered through the streaked, filmy glass at the twilight alley.

"Any activity?" he whispered.

"Nix," came the reply.

The loading docks were empty. Nothing lived in this region of concrete and stone, not even a stray dog or a drunk looking for a place to make a solitary toast to his troubles. Even in ordinary circumstances, it would have been inhospitable—barren and cold. But now, with opposing armies arrayed on either side, unseen and afraid, it was a no-man's-land.

Then a pair of headlights slashed into the gloom and a tiny car nosed its way forward. It stopped. The interior light flashed on briefly. Then the car pulled away, leaving Birch standing there alone. The sight of him moving furtively toward the warehouse was strange because it was real. All the carefully arranged frauds came down to this one small man hugging the walls as he inched forward to the danger.

When the door of the warehouse swung open, Harper caught a glimpse of his enemy. Kerzhentseff's thin, darting form was in the doorway. He was a rodent who had eluded every trap, known only by his droppings and other mischief. Unseen, he had been a presence, otherworldly. But now he was cornered, caught out in the light, his useless arm limp at his side. And Harper could not help feeling that his instinct would be to recognize the danger and to attack.

A hand tapped Harper's shoulder, and he flinched.

"Show's starting," said Braybeck. Harper followed him back into the inner room.

One of the soundmen handed him an absurdly large pair of earphones. Padded against his head, they cut off all sound except for the buzz of his own conscience and a tiny, derisive hiss that seemed to isolate him for scorn. Then Braybeck gave a signal. The technician flipped his switches, and Harper heard Birch's voice.

". . . been here before."

And then, finally, came the other man's voice saying, "Perhaps you are beginning to feel at home."

The sound was reedy, distant, more animal than human. It was the voice of menace, the thin voice of blood. Harper clasped the awkward earphones with both hands tighter to his head. He wanted to be closer, to know this voice as he knew his own, to hear if any note of recognition darkened it. But it only seemed to recede as he strained for it. And in the foreground came a slow, steady thumping that counted off the time like the pulse at the leading edge of a dream.

"What's that?" he asked, his voice roaring in his head. Braybeck's lips moved, but Harper could not hear what they said. He pulled away one of the phones.

"What?"

"Water dripping," said Braybeck. "Damned leaky pipe or something. It'll drive us bats by the time the night's over. Like a broken record."

The thin voice was back in the earphones saying, "Have you had any progress?"

Harper nestled the things back over his ears and listened.

"I've received a medal," said Birch. The same disarming charm, the youth. Now it was turned to irony, and Kerzhentseff was to share the joke.

"I'm pleased for you," he said.

"It means they don't know. I had my picture in the camp paper."

"Is that what it means?"

"They wouldn't give me a medal if they suspected."

"I think they might be that cunning, Birch," said Kerzhentseff.

Harper tried to concentrate beyond the slow tapping, to put it out of mind like the passing of time. He listened to Kerzhentseff going for the sensitive spot, the fear. And he wished he could whisper in Birch's ear, as Birch was whispering in his. But the pulse went on, steady and disconcerting, and Birch was entirely on his own.

"Do you think they know? I mean, do you have some reason?"

"Just my general sense of their skills, Birch. Nothing more."

"I mean, I snatched some papers for you. Maybe I should have been more careful."

"Documents?"

"From the CO's desk. He should have secured them in the safe for the weekend. It was a Sunday. I had duty."

"Is it your duty to rummage through your superior officer's things?"

"The duty NCO is responsible for security, yes. They take that very seriously in this unit."

"Yes," said Kerzhentseff slowly, "and you are a very responsible fellow." Even the static on the earphones, the hollow distance, and the pulse did not obscure the viciousness. "What exactly did you do?"

"I locked up the material and reported the incident."

"But not before making copies."

"The documents didn't mean anything to me. But how would I know? You've been pushing for more, and this is what I came up with."

"Resourceful of you, Birch."

"Did I do something wrong?"

"You have done everything right. Just so. Didn't it strike you, though, how convenient your commanding officer made it for us?"

"He was upset."

"I should think so. Such a foolish mistake for someone so used to handling classified material, and in an organization that takes such things so very seriously."

"He said he wasn't thinking. It really bothered him. It could cost him a promotion."

"Or get him one," said Kerzhentseff. "Let me see the documents."

The dripping water counted off the silence. This was the final stroke. By setting it up so directly, the station chief had made sure that Kerzhentseff would see the possibility he was being tricked. But the trick was so crude that a penetrating mind with due respect for his adversaries would have to conclude that it was too graceless to have been designed. But would the Soviet go further and see that in fact the design was meant to imitate the obvious and that we were clever enough to hide in full view? At each step the reasoning grew more precarious. And at some point no mind could be sure.

"Fascinating papers, Birch," said Kerzhentseff at last. "They speak for themselves. But they do not speak quite so eloquently for their origin."

"I don't understand."

"You may find it unpleasant, but I am going to have to put you on a lie detector, Birch."

"Why?"

"Don't be shocked. Surely you do not expect me to trust your word as a gentleman, knowing you as I do."

"Haven't I done what you asked?"

"You are a liar," said Kerzhentseff. "We both know that. The only question is for whom."

"But what have I done wrong?"

"Perhaps things have come too easily to you."

"You mean I've been set up."

"Well, you might like us to think so."

And then Birch diverted from the text, and it was so dramatic and dangerous that everyone in the listening post turned to Harper as if for an explanation or a decision to abort.

"It's the whores. Nowicki's whores. They must have talked. What am I going to do now?"

"They did not talk, Birch. And they won't in the future. You need not concern yourself. We have disposed of the problem. I thought this had already been explained."

"What do you mean by 'disposed'?"

"You have nothing to fear on their account."

"What did you do to them?"

"I assume you would prefer not to know. You may simply think of them as being out of the picture."

"My God."

"You should be relieved. You see, I pay heed to your fears. They were rational ones, and we took the appropriate steps. You are worth it to us—or were. But now you need to prove yourself. For this I will have to ask some others to join us."

"You've had them killed."

"They are not worth mourning. Nowicki had no taste. You loathed them. Put it out of mind. You have other things to fear just now. This test is not one you can afford to fail."

The voice was slow and soothing now, caressing the evil words. Harper nodded, a signal to let the operation continue. He did not know why, but he was confident that Birch knew precisely what he was doing. Sometimes Birch seemed able to perform as he did by literally remaking the truth. He created a situation in which his natural reactions were themselves, in a sense, duplicitous. The risk he was taking this time was enormous. Harper had seen the tempest the killings had set off in him. But Harper's instincts told him to let the thing go on.

"You're putting that too tight," said Birch. "It hurts me."

"It binds," said Kerzhentseff. "It is meant to bind. Now you must always answer my questions yes or no. Do you understand?"

"Yes."

"Good," said Kerzhentseff. "Perhaps you have done this before."

"Done what?"

"Never mind," said the Russian. "Is your name Jerry Birch?"

Harper nodded his head and whispered the answer, a profession of faith.

"You know that," said Birch.

"Yes or no."

"You call me by my name. I don't see why you have to ask me about it."

"Birch," Kerzhentseff said sternly.

"Yes."

"This is not for amusement."

"No," said Birch.

Harper could not help smiling at his innocent, taunting pluck. But at the same time he worried that Birch was toying with the Soviet because he was afraid. When you are in trouble, buy time. Laugh. The machine cannot appreciate irony.

These were the rules he had taught Birch. This was the technique for trouble.

Q. Is your name Jerry Birch?

A. Yes.

Q. Are you stationed at Camp Zama?

A. Yes.

Q. Is your unit the Third Joint Security Detachment?

A. Yes.

Q. Do you have a security clearance with the code word Black Body?

A. Yes.

Q. Have you ever mentioned to anyone the fact that we have met?

"I'm not nuts," said Birch.

"Birch," said Kerzhentseff. There was violence in his impatience. Don't push him. Play him. Play him.

Q. Have you ever mentioned to anyone the fact that you have been in contact with me?

A. No.

Q. With Nowicki?

A. Yes.

Q. Did you mention Nowicki to anyone other than your wife?

A. No, sir.

A third voice muttered something foreign and inaudible, one of the technicians, Harper assumed. And Kerzhentseff said, "*Khorosho.* . . . For a man who was so upset about the fate of a few foolish women, you are now very relaxed."

"It's so unreal," said Birch.

"You are wrong about that," said Kerzhentseff.

He's trying to give the machines an edge, Harper thought. Be clever, Birch. Make those circuits purr.

"We shall continue," said Kerzhentseff.

Q. Last Sunday, did you have duty at the detachment?

A. Yes.

Q. Were you alone in the office?

A. Most of the day.

Q. In the morning were you alone? Yes or no?

A. Yes.

Q. Did your orders require you to inspect the premises?

A. Yes.

Q. Did you find anything out of place?

A. Yes.

Q. Was a safe open?

A. No.

Q. Did you find any classified documents that were not locked up in the safe?

A. Yes.

Q. Did you find them in the commanding officer's desk?

A. Yes.

Q. Were they provided to you by an American agent?

A. No.

Q. Were they fastened to a file jacket?

A. No.

Q. Do you have any reason to believe the documents are fraudulent?

A. No.

Q. Did you make copies of the documents?

A. Yes.

Q. Did you return the originals to the safe?

A. Yes.

Q. Did you then report the incident?

A. Yes.

Q. Did you keep the copies on your person?

A. No.

Q. Did you put them in your desk?

"You aren't going to guess," said Birch. "The security officer certainly didn't."

"You were far, far too clever for them," said Kerzhentseff.

"I put them in the burnbag until I was alone again."

"They might have been destroyed."

"It was a risk," said Birch. "But I was afraid."

"What if they had been found? What would you have done then?"

"I would have said that I first thought I should make a record of what I found, then thought better of it."

"It isn't a very good lie, Birch," said Kerzhentseff. Not nearly up to the usual standards, thought Harper. Why the embellishment? What did Birch think he was doing?

"It was all I could think of at the time," said Birch. "It doesn't matter. Nobody looked into the bag."

"I want to go through this for yes or no," said Kerzhentseff.

Q. Did you recover the copies from the burnbag?

A. Yes.

Q. And did you take them home with you?

A. Yes.

Q. Are these the copies?

A. Yes.

Q. All the copies?

A. Yes.

A third voice mumbled something, and there was a scraping of chairs. The pulsing water drops interrupted the silence and rattled Harper like an Asian torture. Even though he had penetrated the Soviet's sanctuary, he felt blind and helpless. The machines, whether they gathered the secret whispers or charted the hidden changes in the mind, never told you all that you want to know of them. You were always blind.

"You are lying, Birch," said Kerzhentseff, calmly. Harper grabbed the edge of the table and held it tightly. Birch could not have stumbled on such a simple matter. Were they all the copies? Yes, of course, they were. In truth they were. Birch had received no more. They had not discussed any other papers. What could have gone wrong?

"It is so foolish of you to try such a thing," said Kerzhentseff, sounding weary, weary with the violence he was going to have to do.

"I burned three documents when I got home," said Birch.

"Burned them?"

This was not something they had rehearsed. Birch was improvising. And he was only getting himself in deeper.

"I got scared," said Birch. "I didn't want to go through with it. I didn't want to have them around. I panicked."

"But what stopped you from destroying them all?"

"Donna came home and I had to stop. I hid the rest of the

papers in a book. By the time I had another opportunity I had gotten myself together. It didn't seem so risky anymore."

"You thought about your responsibilities," said Kerzhentseff.

"I guess," said Birch. "I should have told you right off. But I was embarrassed. I didn't want to seem like a coward."

Harper relaxed his grip and settled back in his chair. Birch had done it to show Kerzhentseff the way a lie would look so that the rest would be taken as truth. He was absolutely in control, always had been. He was a master, and the game was ours.

"I don't care about cowardice," said Kerzhentseff, "only about who is the object of the greatest fear. Remember, Birch, we are quite willing to destroy."

"Like you did the whores?" said Birch.

"Less humanely, I think," said Kerzhentseff. "With them we had no quarrel."

20

Within days, Black Body emanations began blinking out all over the world. The Russians were forced back into personal contacts with their illegals. Reports of increased activity came in from everywhere, some fools in the field even claiming personal credit for the change. Dead drops in foul public toilets, furtive encounters in the night, the action was back where it belonged.

It was remarkable, really, the way everything had come together: the hints and rumors about the penetration of Black Body, the good fortune of having an asset so opaque to the polygraph, the final direct confirmation to Kerzhentseff that was too bold not to be true. We had no way of knowing for certain, but we seem even to have scored on the give and go play in which we passed some documents to a suspected Soviet illegal, waited until he had time to communicate the information via Black Body, then scrambled around as obviously as we dared, going through the motions of damage limitation.

In his after-action report, Harper recognized the need to confront the question this success inevitably raised. Was it reasonable to believe that Kerzhentseff had been so thoroughly deceived? This doubt can penetrate the greatest exhilaration; in effect, it is part of the fabric of victory. But Harper argued effectively against it. No single element of the stratagem, he wrote, would have been enough. The Soviets are as wary of the polygraph as we are. Birch's character and low rank would have put Kerzhentseff on his guard, though the same qualities may also have reassured him, since he would doubt that we would entrust to such a man a great mission. The bits of evidence we sowed in various theaters about Black Body's compromise were not themselves enough to raise the alarm. But together all of these devices drove Kerzhentseff into action. It was a simple matter of weighing the risks, Harper wrote. Even if Kerzhentseff harbored his own private reservations about whether he had been tricked, he had clearly done what we wanted him to. It should be of little concern to us, Harper wrote, to know the mysteries of his mind. We had set about to manipulate his behavior, and our success in this was unequivocal. A complete explanation offered itself: since our penetration of Black Body would have put the entire Soviet deep-cover network in jeopardy, the reasonable doubt we had created was quite enough to explain Kerzhentseff's decision to shut Black Body down. We should not indulge in the kind of thinking, Harper concluded, that makes success raise a presumption of failure. Harper's report carried the day, and he was proud that he had silenced the Mocking God as completely as he had silenced Black Body.

Meantime, the dummy detachment remained in operation for a credible period, then folded quietly. Birch was rewarded with a sizable cash bonus, and when enough time had passed that it would not have appeared linked to the operation, he received a promotion. We had him transferred back to a safe billet near Washington, D.C.

Harper's next assignment was to Nairobi. He chose it. He could have had anything he wanted. Even in the sneering corridors of Langley, for a time he had no detractors. A man can glide for a long time on the strength of one success of the order of Black Body. Harper's star was on the rise. And though over the years Black Body was replaced by another system, this did not cast a shadow over Harper's accomplishment. He was verging on a top leadership position in the Agency, which is why he was assigned to a term in internal inspections. In Tokyo he had bought us precious time, which is all that intelligence can ever purchase.

We had very little recorded contact with Birch between the time he returned to the States and the night Harper spotted him outside the Hay-Adams. He suffered no more than the usual problems of readjustment coming home. Espionage is a strong stimulant, and withdrawal is rarely easy. He talked with our psychiatrists, who reported that his symptoms of ennui were nothing out of the ordinary.

Birch offered on several occasions to undertake new assignments, but the files show that he was turned down. At first this was because of the delicacy of Black Body. The Soviets had severed connection with him, and we had no intention of exposing him gratuitously to new trials that might have resulted in doubts about his bona fides. Later, when the urgency of that reason faded, it was decided that despite his special talents, he was not terribly promising as an ongoing asset. He was too ordinary, too squeamish. As one evaluation put it: "These days it is not possible to rely on a patriot who blushes at the sight of a whore." Scribbled in the margin of that memorandum were the words: "He'd go red in the face every time he met his case agent." I didn't recognize the handwriting and frankly never bothered to pursue it.

Of course, we notified the Sisters when Birch came home, with advice that he be left alone and a proviso reserving the right of consultation in any further decisions about his case. Military intelligence also had a copy of his history. It flattered him. While we cannot sing your heroes in public, we do try to make sure that they are appreciated in our own house. Birch's military career, within the confines of what an enlisted man can expect, looked very bright.

Nonetheless, I think it is fair to say that when Harper made his discovery at the Hay-Adams, the surprise at Langley did not go deep. Nothing is ever finally put to rest. We are at all times ready to revise the past. For us, nostalgia always wears a frown.

At its most radical level, this means confronting the possibility that the other side gave us the game the way a poolroom hustler might, to lure us into some larger defeat. The greater the seeming triumph, the greater the potential disaster. I have always felt somewhat sympathetic with those troubled minds who believe that the engine of international circumstance is driven by a conspiracy of all the faceless intelligence agencies of the world. Not that I share this delusion. Having seen the clash of ambition within one agency, I know that the idea of such vast, malign cooperation is absurd. And I have been in the war too long to believe that its purpose is other than to destroy. But the temptation to believe that whatever is not seen must be hidden is always enormously seductive—though only occasionally justified.

THREE
Washington, 1978

21

Instead of turning left at the corner for a straight commute to Langley, Harper suddenly swung left down a long hill into the anonymous knots of traffic. The sun coming up over the low, unattractive buildings did not burn off the haze. At the last moment, he took a sharp left into a side street and checked his mirror to see if any others had beaten the oncoming rush. None had.

He could not have given an articulate reason for evasion. Evade whom? But he had had a taste of the past the night before on the Sisters' stakeout at the Hay-Adams. He was an outside man again. And slipping surveillance was like a forgotten tic that comes back after years of calm. It did not pass Harper's notice that every other time he had occasion to fall back into this habit of stealth, eventually somebody died.

Washington awakened earlier than most cities, and already the sidewalks were busy with the first-shift bureaucrats and their pilot fish from the law firms and lobby groups. All these ordinary people, making government the way others made goods, most preferring not to know what was done at Langley to meet the first obligation of sovereignty, which was to survive. When Harper reached Seventeenth Street he turned toward the mall and then, as if by impulse, swung left onto H. A carryout was doing a brisk business of coffee and buns; the cars were queued up eight deep at a parking garage; the Hay-Adams stood gray in its oblivion of what, if anything, had occurred there the night before. In the little alley where Birch had paused, a food truck unloaded crates of vegetables onto a loading dock.

Harper remembered his last meeting with Birch in Japan. He had taken the station chief with him to Camp Zama. It was all the ceremony they could risk. They had found Birch waiting alone in the quonset hut of the dummy detachment, sitting patiently in the big stuffed chair where he had so often undergone the flutter drill. After Harper had introduced the chief and the chief had made an awkward little speech about the satisfactions of silent accomplishment, Harper had shaken Birch's hand and told him he was sorry the team had to be broken.

"We did it, Mr. Harper, didn't we?" said Birch. "We really did it."

"You did, Jerry," said Harper. "You made the score."

Then Birch did something quite odd and touching. He stepped up and embraced Harper like a brother. Then he whispered the code word and said, "Just give me the word, Mr. Harper, and I'll be ready."

When they left him there and got back into the embassy car to return to Tokyo, the station chief turned to Harper and said, "Strange soul, that one. Dependent type. Makes a good soldier, I suppose. I imagine you're glad to be rid of him." And although the sadness of completion had obscured it then and forgetfulness later, now Harper finally realized that the station chief had been right.

Harper cut back toward the mall. Lines of buses lumbered in file. The big doors of all the stone government buildings were open now, and the employees fished for their identification in coat pockets and purses as they entered. The flags had all been raised, bright against the colorless sky. The sound of traffic was the only anthem.

When Harper finally reached Langley, the guard at the gate

greeted him by name but still wanted to see his badge. Harper parked in his usual spot and slipped on his jacket over the sweaty shirt. He hung the ID from its little chrome clasp on his breast pocket, and when he hit the front door the cold, conditioned air struck him like a blow. The security man inside checked his face against the photo and waved him through the narrows and into the big modern lobby. We are pathological about proving our identities at Langley. We must never walk the halls without displaying ourselves recapitulated on our chests, as if the correspondence between face and photograph had anything at all to do with who a man might really be.

The secretaries were already ticking away when he reached the suite of windowless offices where he worked. Miss Lutkin had already dialed the combination on his door and opened it, even though he had not told her that he planned to come in.

"Am I that predictable?" he asked.

"We called," said Miss Lutkin. "A woman answered and said you were on your way. She didn't sound like your wife."

"Family friend," he said.

"The Deputy wants to see you," said Miss Lutkin, a rebuke. "Shall I tell him you are here?"

"Some dictation for you first."

"He made it clear he did not mean tomorrow," she said, tightening the knot on the prim, unflattering bow at her high collar. Not a glimpse of warm flesh from Pamela Lutkin. She was a secret nobody was out to compromise.

"This won't be long," said Harper. "You limber?"

"I manage," she said.

"Yes, you do. You surely do."

She followed him properly into his office, and as he stripped off his coat, he began reciting the memorandum. He used the words he had been rehearsing since the moment he saw Birch in the alley. Flat, unemotional words that sounded a little better today than they had the nervous night before. She took it down standing, the punishment he meted out for her small insolence. The back of his shirt stuck to his skin as he stood facing away from her, giving her time to catch up with the rush of

sentences. The sweat would no doubt offend her; she did not
believe in it. He had almost finished the report when the telephone rang. She answered it at his desk.

"Yes, he is," she said. "I certainly will."

A bloom of satisfaction spread across her face beneath the
chaste, white powder, and she hung up the receiver and informed him that the Deputy was tired of waiting. He smiled,
conceding her little victory, and dismissed her. Back in his
limp jacket, tie hiked up uncomfortably against his chafed
throat, he went to the Deputy Director's office up one floor at
the other end of the corridor.

A secretary stood sentinel at the inner door. A mark of hierarchy, the door bore two combination locks, and it was hung
with a bright sign that said OPEN in flat contradiction of the
senses. The secretary got up, glanced through the tiny peephole, then buzzed her boss.

"You can go in," she told Harper.

The Deputy had a simple office, no ostentation, no self-
promotional photos of handshakes with Presidents and princes,
no framed commendations. Just a few prints from the National
Gallery, the same Monets and Renoirs you saw all over town
in the offices of men who didn't care. He was a military man,
an Air Force lieutenant general on loan to the Agency. Sometimes he wore his uniform, sans medals. But today he dressed
in civilian clothes, a starched white shirt and nondescript tie.
He could have been an auto salesman.

"Just get in?" he said as Harper stepped inside.

"Miserable heat."

The Deputy leaned back in his chair and looked Harper up
and down as if he were doing an inspection. Over the credenza
behind his desk hung the only striking feature in the place: a
three-dimensional relief map of the world, land and sea, moun-
tain and trough. Three lights on a track shone upon it, so there
were no shadows anywhere.

"Had some excitement last night," he said.

"I came in to write it up," said Harper.

"I hope you found your evening interesting," said the Deputy, his slender figures stuffing black flakes of tobacco into the bowl of his pipe. They said the smoke from that pipe was the signal link between his suspicions and the outside world. When it came out white, they said, he had not yet selected the traitor. When it came out black, he had. The Deputy drew the flame of a wooden match into the bowl, and the heavy Latakia wreathed his cropped, balding head blue-gray.

"I give the Sisters a good deal more credit than most people around here do," said Harper.

"Understand you ran into some old friends," said the Deputy, putting as little weight on it as Harper had in the memorandum he had just dictated, and probably for the same sly reason.

Harper nicked his brow and nodded.

"The Sisters seem to do just fine," he said. "I'm surprised they made Birch so quickly."

"Mad as hell that you didn't say anything," said the Deputy.

"I wasn't sure whom I was dealing with."

"Are you sure now?" said the Deputy, laying his pipe gently by the stem in the ashtray where the smoke rose thin and turbulent.

"Somebody must be running him. I assume it is the Sisters. He's theirs now, isn't he?"

"They say they were as surprised as you must have been," he said with a perfect symmetry of ambivalence.

"Could it have been our operation?" Harper asked.

"Too early to tell," said the Deputy. "The Sisters are being careful about it too. They aren't sure whether someone there might have gotten the thing going without reporting it. You can tell they are worried. They sound so emphatic."

"What do our operational people say?"

"The usual cheery evasions," said the Deputy. "Take it through channels. Sources and methods. Proper procedure."

"Pack of lawyers," said Harper.

"Funny you should say so. I got into it last night with Martin. The Sisters' first point of contact was the general counsel's office. It's come to that now. We're hedged around like a land deed. The face the Sisters put on their concern was that some-body over here might have exceeded our charter. The Sisters have a keen sense of territoriality."

"Do you believe they have grounds to complain?"

"Haven't any idea. But it is one possibility."

"For the record, I haven't been in contact with Birch since Tokyo. I didn't even know he was still in the neighborhood."

"That's a comfort, Richard," said the Deputy.

"I'll get a memo to you ASAP."

"I want more than a report, Richard," the Deputy said softly. "We've decided you ought to take this matter straight through. You know the players. And since there is the question of exceeding authority, I suppose it does fall in your court in internal inspections."

"Are we into an official inquiry?"

"The DCI wants to head off a flap. Call it what you like. If you think you need paper on it, draft it up for my signature."

"I suppose we have to contemplate the possibility that Birch is acting on his own account."

"It would be reasonable to."

"That would put it into counterintelligence. Perhaps that's where it belongs. I'm concerned that I might be a little too close to the situation. It is difficult for me to imagine that he's been turned. I recruited him. I ran him. I was there last night. The whole damned thing. It may be better to find a different ferret. The appearances, if nothing else."

"The appearances are always bad," he said, standing up at his chair to let Harper know the discussion was over, rebuttal heard and rejected. "We've all agreed that it ought to be yours. Counterintelligence will give you an assist if you need any technical help. Don't be shy. You have our backing. Take it all the way and report only to me. And Richard . . ."

"Yes," said Harper as he reached the door.

"Welcome to the NFL."

It was a marvelous joke, really. All of it. A joke on Harper, the newest ferret. At least in the National Football League you could tell who was out to hurt you by the color of their jerseys.

When he got back to his office he delighted Miss Lutkin by curtly telling her he did not want to be disturbed. Then he closed the door and read over the flat words of the memorandum she had already typed letter-perfect, left on his desk, and undoubtedly already copied for her own shadow files. The document was all wrong now; it sounded not simply businesslike and routine but downright indifferent. There's a difference between detachment and lack of concern. The damned paper had, in a matter of a few minutes, turned incriminating.

The phone rang outside and flashed on his panel as he sat rewriting the memorandum in terms more extensive and grave. He could hear Miss Lutkin's muffled voice as she dismissed each call. Mr. Harper is in a meeting. Mr. Harper is with someone now. Shall I ask him to call you when he returns to the office? She loved it, and when he was finished, he buzzed her and said, "This has to be redone," as if the problem had been in the transcription. But before she had a chance to protest, he added, "First you'd better get me Watling at the Bureau. Use the green phone if you will."

"Of course," she said.

The green phone pleased her. It made her part of a secret society, complete with stylized rituals and memorized code numbers. And if even on a clear day a call across town on the encrypting device sounded as if it had been patched through Katmandu to Kabul, all the better.

The amber light on Harper's extension flickered as she dialed the ten-digit code. The scrambler mechanism sat silently in the corner of his office, roughly the size and shape of a refrigerator. Every week an anonymous team ousted him from his desk for a few minutes while they opened the device, changed the dials to new settings, and replaced the tape. It was the kind of rite that could give Miss Lutkin goose bumps: bright, bloodless, aseptic, but utterly exclusive.

The rule of the office was that, since she was senior, no one was to dial or answer the green phone when she was present. As a grudging concession to human functions that occasionally took her to the lunchroom or the ladies', she instructed each new girl on its proper use. One in particular became curious about how the clever thing worked. Miss Lutkin muttered something vague about how it translated the words into nonsense so that anyone tapped into the line would hear only gibberish.

"You mean that if I said, 'The pen is on the table,'" said the girl, "it would go out over the line as 'My aunt has green thighs'?"

"Please," said Miss Lutkin.

Harper could not help himself. "That's exactly how it works," he said. "A lie box."

She had a cute laugh, short and ticklish. He knew she would not last.

Miss Lutkin appeared in the doorway. She preferred not to debase the green phone by referring to it on a common intercom.

"I have Mr. Watling," she said. "I had to use Priority Two."

"Very daring," said Harper.

She did not dislike him particularly, he was certain. She disliked him generally. She had an aversion to all outside men who came inside. They never quite washed clean of their hidden pasts, full of imagined exploits and risks, all unsavory. It offended her sense of order.

"Watling?" said Harper loudly into the hissing receiver. "Just trying to get hold of all the string before setting out to unravel it."

"Bad start," said the shrunken voice. "You did a number on us."

"Not at all," said Harper. "I just wasn't sure of the players."

"How come I have the feeling I'm being set up to take a dive?"

"I know what you mean," said Harper. "They put me on the case."

"Maybe you *are* the case," said Watling.

"You ever do time in the organized crime section? Did you ever deal with the hoodlum element, Watling?"

"What I did and what I didn't doesn't have a damned thing

to do with it. Except maybe what I didn't do when I realized what you didn't do last night. But don't think I didn't want to do it."

"It's a theory I have," said Harper. "People take on the characteristics of their enemies. Growing together over the years, like man and wife, master and dog."

"I don't like dogs," said Watling.

"Maybe we'd better get together and talk," said Harper, "starting with what you folks were doing with my guy."

"See, that's your problem," said Watling. "He isn't your guy anymore."

"Bad habit."

"They're not going to like it much on the Hill. They think we have a law about these things."

"I suppose you've already notified them."

"Hey," said Watling, "I sees my duty and I does it. I'm not going to be the guy holding the turd when the Chairman reads about it some morning in the *Post* over his cornflakes. You weren't around here during the purges, were you?"

"The newspapers onto this already?"

"Give them time," said Watling.

"I take it that you are fairly confident that the Kerzhentseff drop wasn't your show," said Harper. "My guy never does more than he is told."

"There you go again," said Watling. "You have to get with the program. Say it this way, Harper, 'The FBI's asset.' Say it that way just once to try to get the hang of it."

"The FBI's asset, Senator, seems to have gotten away from them."

"Shit," said Watling, and Harper couldn't help thinking of the girl's ticklish laugh and wondering whether that word went out on the line as "Sisters."

"Either you don't know this town, my friend," Watling continued, "or else you are one very hard guy. Or then there's the possibility that you are just plain stupid. See, somebody somewhere is going to be in a mess, and it's going to be so deep he won't be able to see out of it. And I'm only really sure of one thing: it won't be me. I were you, I'd be a little concerned where I was going right now."

"You going to be in tomorrow morning?"

"All day."

"I'll be there first thing," said Harper.

"Wear cast-iron slacks," said Watling. "I got some guys down here want to bore you a brand-new asshole."

22

The house was dark when he returned that evening. The dead-bolt lock had been thrown closed. He was surprised that Fran was not there, but not particularly disappointed. He had spent the afternoon reading over his files from Tokyo, and they had taken him back not only to Birch and Black Body but also to the first, unbetrayed years with Janet.

A gap of several days in his memoranda represented a trip they had taken to the snow country, the traditional inn where they had stayed in a room so tiny and fragile they were at first afraid to make love in it. He remembered the tearoom near their apartment where they used to go for dinner. Every table was private, screened around with rice-paper walls and set in a pit in the floor in such a way that it simply invited the diners to touch toes and laugh.

He had come across his cable on the Saigon disaster, and it brought back to him the time when they had met. She had been in Saigon working on a contract study of bomb damage to the agricultural economy for USAID, and remarkable as it seemed for someone who had come of age at Smith in the late sixties, she had no strong opinions on the war.

"I'm not disillusioned," she had told him once, "because I guess I was never really innocent."

"You mean I'm not the first?" he had said. At the beginning, they had been Nichols and May in the combat zone.

But then came Harper's blooding. She had stayed with him, taken him home to her flat when the ghosts in his own were too much to bear. It was the decent thing to do. You protected your own.

"It's a horrid war," she had said in the only judgment she ever made on the subject, "a horrid, horrid war that does things like this to people like us. It just isn't worth it." And he had never heard a franker statement of why we eventually cut losses and ran.

After his transfer to Tokyo, they got together in neutral territory. Taipei, Macao, Hong Kong, Singapore. He was bored with work, and there was no urgency between them. Then came Black Body, the uncertainty. They met once in Singapore for a holiday, a week at the Raffles, the garlanded pool in the courtyard, the Chinese waiters in white, dinners on linen. Order was there for the asking. He rediscovered how much he needed her.

But it had not really been the same since Tokyo. Not that anything had gone bad, but the sweetness of order became a kind of routine. She had resigned the Institute to go with him to Nairobi. To her surprise, it had been impossible for her to find decent work there. The private firms were full up; the UN was nervous about too many Americans. Harper's cover with the Embassy closed off that possibility as well. She had to play the dutiful foreign service officer's wife, and she did it quite persuasively, exacting from him in return only a promise that he tolerate the demands of her career next time as quietly as she tolerated this.

They settled into a kind of numb comfort. But they had never been drawn to one another for passion in the first place. So when life and loving lapsed into habit, neither of them showed resentment or surprise. And when they went on an exhilarating trip into the game reserves, or when the right mixture of alcohol and heat put them into a kind of rapture, they took it as a gift, a sunset on the Serengeti.

But around them the so-called sexual revolution was moving into its reign of terror. They saw its manifestations among their friends. It was a revolution of rising expectations, and Harper liked to think of himself, good husband, as its Edmund Burke.

One evening they were trapped into a dinner party given by the local representative of an American bank and his wife. It was too formal, and much too frank. The hostess made it clear that she did not care for Africa or Africans. Harper was appalled that she said it in front of the black help. And under the influence of too many glasses of serum, she made no effort to hide her disenchantment with the host.

"I thought I would be filthy rich with him," she said, "but all I am is filthy."

"You married integrity," said the host, pouring himself a defense.

"Who would have known?" she said. "Harvard, Chicago Business. You graduated so well."

"I feel fortunate," said Janet, courageously, "to have the opportunity to travel."

"Seven varieties of dysentery," said the hostess as she inspected a cracker. "If you've had one, you've had them all—and we have, haven't we, Harvey?"

"Margaret," said the host, "would prefer not to have to shit."

It was dreadful, and they left before the cigars had burned halfway down. When they got home, Harper brought up the Birches, how different they had been, how much simpler and more natural. They traded off the highs against the lows, he said, gave up the adrenaline illusion.

"Is it adrenaline, darling?" Janet said. "I always thought it was testosterone."

"We could be like the Birches," said Harper.

"They seem awfully dreary," she said, "and you sound so patronizing. Are you almost finished with your wine?"

Harper passed through the house in Glover Park looking for

traces of Fran. A hairpin, perhaps, or a tissue smudged with makeup, some piece of clothing giving off her sweet, yeasty smell. She was there, somehow, a disorienting field of force, just as Janet was, in absence. Until the night before, the two women had stayed in their separate spheres. Janet was Glover Park: predictable and proper. Fran was Alexandria: exotic and sinful as the name. He wanted them both. He wanted the polite and comfortable intervals with his wife. The space between them never varied. They were constant. He also wanted the intensity of Fran, heightened by the impermanence she insisted on. And sometimes he believed that, more than either, he wanted what the Birches had, the closeness that came so naturally to them.

Now all these conflicting desires had come together as presences in the room. And the interference pattern was as strange as the Escher prints Janet liked so much, two hands holding pencils, each drawing the other. You could not locate yourself in it, could not tell who was imagined, who revealed.

Janet had been delighted when Harper had received his posting back to Langley. She had never said she was bored in Nairobi, but he knew that his unsharable life was not enough to carry them both. As soon as they returned, she was off to the West Coast to try to pick up her career again. Harper handled the settling in alone. He was pleased with the house he had found for them. The back room was perfect for her study. The side bedroom, once it had been fixed up and fitted out with shelves, served passably as his reading room. He liked the age of the place, the fireplaces and their marble hearths, the dark wood of the floors worn down in patterns which, hidden by stain and varnish, revealed their years only to the familiar touch of the toe.

Janet's work involved travel, and Harper was mired at his desk. She was happy, and when she wasn't on the road for the Institute, he shared her satisfaction. Still, when she was gone and Harper returned home from a day of eyestrain and cables to an evening of port wine and silence in Glover Park, he found himself often thinking of Fran; and when he had finally called her, she had simply asked him why he had waited so long.

Infidelity, to Harper's surprise, had come quite easily. He wanted to find this distasteful and sometimes convinced himself that it was, the way any mastery can be when it sets one apart. Janet was too busy to distrust him. She expected things to be held secret from her. This was the nature of his work. And it made it just that much easier for her to keep the proper distance.

But Fran . . . Fran wanted everything, especially what he denied her. Harper assumed that she would not have stayed with him were it not for that touch of darkness. One weekend in Williamsburg, in bed, with a warm, darting fire in the fireplace, he had told her about Vietnam and the dead.

"It must have been wonderfully scary," she said.

"I could have done without it."

"Is that where you met your wife?"

"She didn't believe in heroes either. We were a natural."

"If I'd been there I'd have whisked you away in a boat. Like Catherine in *A Farewell to Arms*."

"I wouldn't have gone, you know."

"Don't be so sure. I can be very persuasive."

"I didn't think you let yourself want anything so badly."

"I can be a good liar too, you know."

Harper could not find in the house any sign of Fran's visit. She had left without a trace. He made himself a salad in the kitchen and poured himself some ordinary wine. The food did not interest him, and the wine did not soothe. He dialed California, but Janet was not in her room at the hotel. It was still early there, and he left a message that he would try again the next day. Then he pushed the receiver button down, slowly, carefully, as if he were clutching some piece of heavy machinery into reverse, and let it rise against his finger before dialing again. Fran answered the phone on the first ring.

"I was wondering," he said, "whether you would want to come and terrify me again."

"Maybe it would be more exciting for you if I refused," she said.

"You didn't leave anything behind for me to remember you by. Have you had dinner? We could go somewhere."

"Somewhere dark and anonymous," she said. "Somewhere that nobody *ever* goes to be seen."

"I think I know the place you're talking about," he said.

"I've eaten," she said. "But what if I just rushed to your side in desperate longing. Would that be all right?"

"We could try it."

"We really should take full advantage," she said. "There will be time enough to be furtive when she comes home."

He heard the edge on it and took it as a kind of flattery.

"I'll run out and get something," he said. "Champagne do you think?"

"To celebrate?"

"Our luck."

"And our conspiracy," she said.

When she arrived she was dressed in a mannish shirt and jeans. Harper was still in his suit.

"You needn't have come formal," he said.

"I bet you wear a tie when you wash your car," she said. "I'm going to have to take you to the Georgetown boutiques someday."

"She would know something was wrong," he said. "Senescence."

His reference to Janet did not go down well. Fran stood in the doorway studying him. Then, as if she had decided he was worth the risk, she shrugged, pushed past him in a rush, and fell into an armchair.

"How could you have a second childhood?" she said. "I don't believe you ever had the first."

"Are you old enough to have a little wine, young lady?"

"Just," she said. "Forever just."

He held the sweating bottle in one hand, carefully undoing the little wire basket that restrained the cork.

"Be bold," she taunted.

"There are two kinds of people in this world," he said, "those who make champagne corks pop and . . ."

"Cowards," she said.

"Secret agents," he said. "It's a question of whether one needs to draw attention to oneself," he went on, holding the bottle at a safe distance as the cork sniffed quietly into his palm.

He poured both glasses without spilling a drop. "It's only California," he said. "The store down the street didn't have anything else cold."

"Commemoration of the one who is absent," she said, sipping it. "A little too sweet, but always available."

He let her have her moment. There was no pleasure for her in his duplicities. She had nothing to lose.

"I thought you might have decided to stay here today," said Harper.

"The way you were acting it didn't exactly look promising."

"I had a problem."

"One of those ugly secrets?"

"Very closely held."

"Why do the words always sound so sexy? Closely held. Is it really any more than paper?"

"Paper," he said and then added, because he knew it would delight her, "just paper and flesh."

"Sometimes I think you are little boys at their play," she said. "Little boys with dirty magazines behind the garage. Paper and flesh."

"I thought I didn't have a childhood."

"This isn't childhood. It's puberty. Perpetual puberty. You take yourself very seriously."

"Merely life and death," he said, "and the survival of freedom as we know it in a hostile world."

"I know all about it. Exploding cigars. Peeping into bedrooms. Experiments with acid. If that isn't adolescence, I don't know what is."

"Those were the bad old days," he said.

"I suppose that's meant to be reassuring."

"Just terribly intriguing," he said. "Spy stuff."

"You're not a spy," she said, warming to the idea.

"So you are meant to believe," he said. "A spy does not advertise. He does not make the champagne cork pop."

"The trouble with lies," she said, delighted, "is that they make you begin to doubt the truth."

He put records on the turntable, the Beethoven quartets. Janet had bought the set and a score to go with it. Sometimes she followed along—a mathematical appreciation, he assumed. But it did not matter now because Fran did not know whose they were and Harper did not care. When he turned back, she had moved to the couch, lounging there with her legs drawn up beneath her. There were times when she was as close and untouchable as a daughter, times when his own feelings horrified him. Then, at other moments, it was she who was teaching him. But even then he wondered, like a parent, where the young had learned so much about danger.

"Was it a bad day then?" she asked.

"Began rather poorly. Ended on a calmer note."

"Leaving our form of government safe for the night."

"Same precarious balance," he said. "But I think it may hold."

"That's a relief," she said, resting her chin in the palm of her hand. He put down his glass and kissed her.

"So is that," he said.

"Everyone has a small part to play."

"Duty?"

"Don't believe in it."

"Very unpatriotic," he said as she drew herself up and pulled his hand melodramatically against her breasts, soft and free under the cotton. It was not exactly an invitation, more like a dare.

"I guess I just go with the winner," she said.

"Will you stay then?"

She kissed him and then pulled back.

"Still too early to tell," she said.

Harper took off his coat and tie and then relaxed back in the couch, so close to her he could feel her heart beating against his arm. The music was slow, insouciant.

"What was your husband like?"

"Like your wife," she said. "Very busy."

"Would it be the same, I wonder, if it weren't for them?"

"You learn the best doesn't last. It's a bummer. But you learn it."

"Are you sure it's always true?"

"Something ends as soon as you start to trust it. I'm sure of that."

"Then we're doomed, you and I."

"I don't know," she said. "I only trust what you don't let me see."

"Things do get turned around. I don't know how it all got started."

"I'm not old enough to remember," she said.

"Will you ever be?"

"Let's don't talk the edge away," she said, perhaps just this once a little afraid. "I don't want to be a problem we can solve."

"Life on the edge."

"That's where you live, isn't it? Maybe that's what I want too."

She pulled him gently onto her and they embraced and kissed in a convincing likeness of passion. Perhaps she was right about trust. Enchantment was a kind of surprise, and belief could not be surprised, except by betrayal. He worked his hand up under her shirt. She moved to make it easier. The flesh was soft, arching against his palm, its point hard and alive. Always that fine paradox: soft and urgent, hard and yielding. It was a small truth, unwilled, incapable of deceit.

When he was young he thought, like all men, this truth was large. He thought it was very nearly everything. The first time, amazed and grateful in the front seat as another, more experienced couple thrashed in the back, he had put his virgin hand there and wanted to yelp for joy when the girl did not squirm away. Up under the stiff, ribbed elastic, the still-growing flesh was sentient, stiffening in the sympathy of mystery and excitement. The elastic came up and off, tangling in his fingers. He worried; did it hurt her? She breathed in his ear, an unintelligible word that meant only that it was all right. They were so very new together, and yet he was sure that when the body rose and stiffened, it was giving its eternal benediction. It was true. It would last forever. What was her name? What had it been?

He opened Fran's shirt and lowered to touch his lips to what he knew. She turned, presenting the other momentary truth, and he took it.

"You are better tonight," she said.

"I'm fine."

She ran her hand up the inside of his thigh.

"Yes you are," she said.

Then they were on their knees, bare to the waist, touching. His fingers found the button of her jeans. It popped open and the fabric parted. Beneath she was warm and naked.

"Evil," he said.

"Ready," she said.

He was arush with discovery, the territory where he had wandered before but which remained always uncharted. He found the deep folds and explored them.

Her breath drew in sharply. Was it unknown to her, too, her own flesh? She shuddered, and then: "My God."

She began to laze back, but he held her up. He wanted them upright, on their knees, devotional. He wanted this touch to praise her. This once he did not want to give way to urgency. There was time; the truth could be made to last, if only for a little while. And when she moved to touch him, he stopped her.

"Not yet," he said. She gasped as if he had found a nerve, and then she began softly yet deeply to laugh.

"Oh my, yes," she said.

But something pierced her laughter, a cry of pain. Harper thought he had hurt her. But when the sound came again, he realized that it was not her at all. It was the telephone. She drew back the slightest distance—to the other side of the universe. He heard the music again, bright and orderly, broken by the telephone's insistence.

"I have to answer it," he said.

He stood up and paused over her for a moment. She slumped there on the floor as if she had been struck. He draped his shirt over his shoulders and went to the kitchen.

"Hello," he said, not graciously.

"What a tone," said the voice. It was Janet. "Did I get you out of bed?"

"I'm half-naked," he said.

"Sorry I'm missing it," she said, offhanded.

He looked back through the doorway. Fran had not moved except to go limp. He stepped around the corner out of the light.

"What was it you wanted?" he said, as neutrally as he could. Even in whispers Fran would hear him. And, as practiced as he was in ambiguity, he wasn't sure there were words that could hold both of them at once.

"There was a note that you had called," said Janet.

"I said that I would."

"Is it late there?"

"The usual interval," he said, managing, he thought, to put a small smile in it.

"I didn't disturb you, did I?"

"It was a hard day," he said and peeked around the corner again. Fran was up on the couch now, dressing, pointedly looking the other way. The records had ended.

"Richard, are you still there?"

"Of course," he said, retreating as far as he could into the dark corner. "I thought I heard something outside."

"I asked you whether the material from New York had come in the mail."

"I'm sorry, Janet," he said. "No it didn't." He had not even thought to look through the stack from the mailbox.

"Too bad," she said. "I was kind of hoping that was why you called."

"No. I just promised I would."

"You don't have to make it sound so odious."

"I don't mean to, Janet. We have our saving little routines."

"Our little secrets."

"I wanted not to disappoint you." He tried to get the tone of it just right, but he heard it shading slightly away from her. She did not pick up on it. The telephone was a curse, but it was also useful in what it concealed.

She said it warmly, or so it seemed: "You never disappoint me, Richard."

And for an instant he imagined her at the other end of the line knowing everything but too proud to reveal that she did. His hand was damp on the receiver. He rushed ahead, perhaps too quickly.

"How long do you think you'll have to be gone?"

"Do you miss me?"

"Very," he said.

"It will be a week at least, I'm afraid. Maybe longer. But it's going very well. Sid says it may be the most valuable project we've done in years. The board is excited. Sometimes it scares me."

"It shouldn't," he said. "Don't let it."

"I get scared because of the whole war thing. We thought that was terribly exciting, too. A great opportunity. A laboratory."

"You did," said Harper.

"That's what I mean. They did too."

"They were fools. They didn't know a damned thing about guns except their muzzle velocities and range. But this time the guns don't come into it. There's no danger."

"Maybe not. It's just that sometimes I think the guns are always there, only hidden."

"The tragic view."

"I'm learning from you, I guess. Original sin."

"Sin isn't all that original."

"Not to you, perhaps," she said, and when she paused he cowered with the sick feeling that she had known all along. But then she said, "Your work is more honest than mine that way." And he knew that she meant no more than what she said.

"Some of my colleagues think I'm a hopeless optimist, you know," he said. "It's all a matter of the company you keep. Enjoy their enthusiasm. Ride with it, Janet. If you're skeptical, keep that too, but keep it to yourself. It puts you that much ahead. You know what can happen."

"Life is politics," she said.

"Politics is politics," he said, which was a way of agreeing.

"I'm glad I called you, Richard. I feel a little less frail now. We do so well at a distance, don't we?"

"On the telephone we are pure."

"No diversions."

"Sadly, none at all."

"I miss you. Will you call tomorrow?"

"Sure."

"I won't talk about danger tomorrow. I'll tell you about flowers and the sea. We're going out to the shore in the afternoon. I'll think of you in the waves. I love you."

"And me you," he said. If she could tell why he did not want to use the words, if she realized it was not just because they were not suited for the wires, she did not show it.

"Sleep well," she said.

"You too."

Then the line spat and went hollow. Far off, another conversation bled onto the circuit, indistinct as the transmission from a faulty mike. He had to strain to get the words. The man seemed to be offering to stop by a store on the way home. The woman was telling him she would go in the morning.

"It's no trouble," he said. "Save you a trip."

"You're tired," she said. "It's all right. We have enough for breakfast."

The poor connection accentuated the rise and fall in their voices. They sang like nesting birds.

"I'll stop anyway," he said.

"You're sweet," she said.

Harper resisted the urge to intrude on the line and offer his blessing, the urge he had so often felt as he sat overhearing the Birches' enviable ordinaries. So, silently, like a god in the machine who does not deign to reveal himself, Harper rang off without approval or blame.

He stayed in the corner a moment more, collecting himself, thinking about what to tell Fran. The trouble was that he had stepped back from the edge now and did not want to return to it right away. You could live in two contradictory worlds, keep them in mind at one and the same time, but even the strong could not always travel between them at will. He took a breath and moved into the trapezoid of light from the living room.

But there was no problem. Fran had gone.

23

Harper's ID had no magic with the guard at the door of the Hoover Building, so he had to wait as the man called upstairs. A couple of young cadets lounged in a glass-walled room flirting with a secretary. When the guard finally assured himself that Harper was safe, he offered to have one of the young men take him up. Harper declined. The guard insisted.

The cadet was taciturn and serious as he led the way down the stark, undecorated corridor. The building was too new to be enriched or tainted by the past. A car was open and empty when they reached the elevator bank. Some young girls got on carrying bags of fast food and soft drinks in wax cups. They chattered about somebody's wedding. Then a middle-aged man in a sweatshirt and shorts squeezed in, still panting from his jog. A secretary was waiting for Harper at the proper floor.

"Mr. Harper?" she said.

"Yes."

"Look, I'm supposed to take you to the conference room, all right?"

"Fine," he said, and she missed a stroke on her chewing gum when he gave her a little bow. The cadet seemed satisfied that the secretary could handle any mischief and excused himself.

"Traffic bad?" said Watling as Harper entered the long, sunless room.

"Not very."

"We've been waiting."

"I didn't realize I'd have to have a security investigation in your lobby."

"Can't be too careful," said Watling. "This is LaSalle. Vitale. Robertson. You already know Malloy."

"I owe you an apology," Harper said to the Irishman.

"You can imagine our surprise," said Malloy, but not meanly.

"LaSalle's the case agent," said Watling, "for Birch I mean."

"Man of the hour," said Harper.

"Time will tell," said Watling. The five of them arrayed themselves along one side of the conference table. Harper sat opposite them like the man in the dock. "LaSalle is new on the case, which we show as pending inactive. His predecessor just moved to the L.A. field office, an ASAC in the intelligence unit. Good job. We put his ass on a plane back this morning, but he'll be awhile arriving."

"What had been your relationship with Birch?" Harper asked.

"File shows nothing more than one routine conversation, laying out the ground rules, giving him a phone number and a name just in case. The usual."

"How about Kerzhentseff?"

"We've watched him, but there hasn't been much to see," said Watling. "That made us a little nervous. A guy like that doesn't just hang around going to parties. But until last night he was clean."

"The file show any other contacts with Birch?"

"None," said Watling. "We didn't have Birch wired, of course. Got to prove a guy is Brezhnev anymore just to get on his line. He was on his honor to report, according to the file."

LaSalle pushed a file jacket across the table, and Harper flipped through the pages, seeing a vague description of his own work with Birch in Tokyo and a few later entries.

"You'll copy me on all this?" said Harper.

"Whatever you want," said Watling. "We'd like to see what you've got too."

"Is everything here?"

"If it isn't, our guy in L.A. is going to become our guy in Butte," said Watling. The others thought that was pretty amusing, and Harper decided he should too.

"Could Birch have been confused about what you wanted him to do?" Harper asked. "Maybe he heard Kerzhentseff was in town again, figured you wanted him covered, got that impression somehow, and made the move."

"He hasn't called us yet," said Watling.

"No wonder you guys are such miracle workers," said LaSalle. "You must treat assets like kids in a Montessori school."

"Don't be tailoring the hair shirt for us quite yet," said Harper.

"You got to start sometime," said Watling. "He ever do that sort of thing before?"

"Go off on his own, you mean?" said Harper. "Not on my watch. Birch was a very straight arrow."

"Maybe you are not entirely objective," said LaSalle.

"And you are, I suppose," said Harper.

"I don't know about the rest of you," said Watling, "but I'm as objective as hell. I'm a regular machine so long as it doesn't look like I'm the guy to take the fall."

The laughter was more than polite. It was frightened and real.

"We have the main action on this right now," said Watling. "The man has been in contact with a known Soviet intelligence agent. He's got a security-clearance job. It's a straight espionage case as far as we're concerned. Unless, of course, you were running him out of Langley. Then it would be a little more interesting."

"I haven't been in contact with him myself for years. I signed off in 1971, and when I got my receipt, there was no damage to the equipment reported. Initially, it does not appear that anyone at Langley had any involvement. That's part of my assignment, to nail that down."

CONVERGENCE

"Our counterparts giving you a hard time?" said Watling.

"No harder than usual."

"If it's the usual, you've got a problem, friend. But that's none of my concern. I'll tell you this, though: When we get our ducks in a line we are going to go hard on this. And if your people aren't ready, we'll run right over you."

"Fair warning," said Harper.

"Fact. The candor option," said Watling. "Here's more: I think we're going to find that your confidence in this character Birch is misplaced. I'll give you odds that he's the one who'll fall."

"That's crossed my mind," said Harper. "I haven't overlooked any of the options."

"I'm telling you to be careful because he can bring you down with him if you don't watch out," said Watling as he rolled a government-issue black pen back and forth between his thick fingers. A fluorescent light began to flicker disconcertingly. The other men were easy in their chairs. But if Harper had chosen the candor option himself he might have put a similar warning to them. The simple fact of the matter was that Birch was in a formal sense theirs and not ours. And if he had gone bad, there would be enough blame to cover them all. He could have told them that, told them that he had every intention to lay the thing off on someone else too. But he didn't want a confrontation.

"You've all been very helpful," was all that he said. And then they shook on it politely, all of them, because it was still too early for alliance or exclusion.

When he reached his office the document he had demanded was waiting for him, eyes only, under its color-coded security coversheet on his desk. Ordinarily, the people on the Agency's operational side balk and quibble when they are asked to go on record about a particular name. Even for purely in-house consumption, it is difficult for them to write a definitive yes or no. For one thing, they worry about the risk of compromise if they acknowledge that an asset was in fact in use. But even asserting that a man was not in play carries a risk. It makes the enemy's process of elimination just that little increment easier, and there is always the possibility that the individual they dismiss has actually come under the control of the other side. It never pays to say anything unless it gives you leverage.

There was still another reason for their reluctance, Harper thought. Fundamentally, they rejected the regime of yes or no. The Agency's operational side was not a binary world. In the nether realm of possibility in which they were cursed to spend their days, the strongest word was *might*.

But when it became clear from the Deputy's direct involvement that the silent partner would taste the lash, the operational side overcame all its scruples and disgorged its testament:

"Jerry Birch: no contact in U.S. subsequent to routine reentry processing. No tasking. No active case agent involvement. No hostile foreign intelligence service approach reported by subject to this Agency, as required per instructions in general field debriefing recorded as administered by Richard Harper, last active case agent, Tokyo 11/17/71. Operational control transferred to Federal Bureau of Investigation pursuant to DCI Order 50.100. File inactive to date."

They did not miss the opportunity to name Harper, nor did he expect them to. The ammunition against him was there in the record, and nobody likes an inspector. He was pleased at first to see that the Agency's own files appeared in order, that Birch was apparently being run by someone else. Of course, he could not simply accept the document on faith. He was obliged to make the rounds, interviewing all the principals about what the record might have hidden. But when the Sisters came back to him with a report that their hands were also clean, he began to see the true nature of his own problem.

The inquiry had narrowed in on Birch and Harper, and Harper was forced to consider what he might have to do to Birch in the course of proving his own lack of complicity. He did not want to do his old friend any harm. But like Birch's experience in the brothel, the not wanting made the doing something he could contemplate. Still he chased other possibilities, ran down phantom leads. But he found himself pursu-

ing an investigation that, the further he drove it, the more it
kept doubling back in his direction. And it was in his increas-
ingly desperate effort to turn the thing outward again that he
finally called on me.

24

As a matter of formality, our discussion was styled as an interview in Harper's inquiry. I agreed to his request that the conversation be between the two of us alone. No scribes or corroborators, no aides or attorneys. It is a new era, and we do not find it uncommon anymore for our personnel to retain counsel when a significant internal investigation moves in their direction. In the past this would have itself been grounds for dismissal. But since the purges, the tradition of team playing has broken down. It is every man for himself.

I believe that Harper was genuinely relieved when I did not insist upon assistance. As it turned out, he was not really interested in facts anyway. He seemed haggard and quite at wits' end when I greeted him at my door and showed him inside. He has never struck me as a man with any great reserves of stamina, and it was obvious that he had been pushing himself to the limit in trying to resolve the curious business with Birch. His eyes were bloodshot—drink, I suppose, to help him sleep. His usual careful appearance was beginning to slide. He needed a haircut, and I noticed that his fingernails were chewed to the quick.

"Sit down, Richard," I said.

"It's good of you to indulge me," he said. "I don't suppose there is very much of a specific nature that you can provide me. You were in the decision loop on Black Body, but since then I gather your connection with Birch has been as remote as my own."

"I am not acquainted with the nature of your connection," I said, but I softened it with a tone of irony I could see that Harper appreciated.

"I am unconnected," Harper said, "very much unconnected. That is my problem. I can't seem to make any linkages at all. Proving a negative, I suppose. Always a problem, especially when it is in the nature of an alibi."

"I am sure it will not be particularly helpful," I said, "but I am prepared also to disavow any contact with Birch."

"Everyone is."

"But you don't believe it."

"I believe that I saw him with Kerzhentseff again and that it is not at all like him to freelance."

"From what I know, I think I would agree," I said. "You know, Kerzhentseff found himself in a spot similar to yours when he returned from Tokyo to Moscow after Black Body. He had discovered what appeared to him to be a major compromise. And it had been in his very own back yard, the system he himself had promoted. I remember the chatter here at the time. You were still in the field then, I believe. But all our thinkers were saying that in addition to disarming their communications security, we had probably managed to do in Zapadnya as well."

"Would that we had," said Harper.

"He did not simply defend himself, Richard. Zapadnya went on the attack. First he scoured the technical services branch of every specialist who had filed a boastful report of Black Body's capabilities. Then he turned his attention to the users of the system. He assumed that the Americans must have had help in breaking the system, and he was ruthless in fixing blame."

"I'm afraid Kerzhentseff was in a bit different position than I am," said Harper. "He had the authority to destroy. It isn't the same system. The standard of proof there is less exacting."

"You've always said we were growing toward one another, the Soviets and us," I reminded him. "Perhaps in this too, if you are willing to entertain the subversive thought."

"I don't see anyone it would be proper to go against even if I could," he said, rather pathetically, I thought. "Do you really suggest that I find a scapegoat?"

"You came here for advice," I said. "I am only telling you to keep an open mind."

"That's the disease, isn't it?" he said. "An open mind."

"I had assumed that you were prepared to be baptized into —what is it that you call it?—the cult of possibility."

"Perhaps I've already been confirmed," he said. "I've let myself imagine all kinds of things. Shall I go through them?"

"If you think it will help," I said. I did not know why he had singled me out as his confessor. But I was not about to turn him away. I will never spurn a man who wants to give me a glimpse of his mind. Espionage is a continuing education in human frailty.

"If Kerzhentseff had controlled Birch in Tokyo, controlled him all along even as we tried to use Birch against him, then why would he have shut Black Body down?"

"Toward some greater end," I said. "That is the usual argument."

"Yes," he said. "And the end is usually a man, someone within our ranks for whom they are willing to sacrifice in order to promote or protect."

"Someone much better placed than Birch," I led him.

"Much better. Someone at my level," he said.

"Or mine," I obliged.

"But correct me if I am wrong; no one at Langley prospered particularly from the success we had against Black Body. It was a collegial operation. Everyone was in, and no one had the lead."

"Other than yourself," I said.

"Of course."

"This is the nub of your problem," I said. "It occurred to me quite early on, I have to admit."

"The other possibility is that Zapadnya wanted to do his own employers harm," said Harper.

"You mean that he was actually in our service, without your knowledge, I presume. Double blind in Tokyo."

"I would know it by now if he had ever been on our rolls," Harper said. "I've been quite thorough, and nobody has held back files. But what if he had been operating entirely on his own command, out of some secret ideal or complaint that had a first claim on his heart? What if he had been on our side without actually being with us?"

"There's that idea of yours again, Richard. You know that it does not appeal to me."

"Too benign," he said.

"Too convenient as an excuse for weakness."

"I know that it won't fly. I can hardly write a report of this affair and say that they, neither of them, meant us any harm. God's in his heaven and all's right with the world."

"The Mocking God?" I said.

"Heaven isn't that one's province."

"What about Birch?" I asked. "Maybe that's the way into this. There's always the chance that he made the first approach to Zapadnya after we showed such an indifference to his offers of continued service."

"I suppose I'm driven to that option," said Harper. "I just have the intuition that it isn't right."

"Trust?" I said.

"I wouldn't put it so baldly," said Harper.

"Best not to."

He stood up and went to the window. It looks out into the trees, a picture of the land as it was when it was innocent. I find the view quite refreshing, even knowing that just below the surface is our war room.

"I suppose I also have to suggest that someone might want to set me up," said Harper.

"That would be the last refuge, I think, unless you have proof."

"Yes," he said. "They would probably think it was desperation, a pathology."

"Many have made that mistaken assessment," I said. And he understood that my point was meant to be at his expense, for since his breakdown in Saigon, it had always been his view that the unseen menace must be treated as unreal. "But I wouldn't be too quick to discard it if I were you. The rendezvous took place at the precise time and place you would be present, after all."

"I've looked into it. The assignment was perfectly random."

"Surely someone knew in advance."

"Anyone might have. The training lists are not closely held."

"The distribution is very wide, I suppose," I said.

"Somebody would have to have been in a position to manipulate both Birch and Kerzhentseff," he said. "Not very likely."

"Not impossible, though," I said. "Don't forget Zapadnya. The secret of his success is an instinct for the improbable, and utter ruthlessness."

"The description of a dangerous man," said Harper.

"Or of one who wants to get out of danger," I said.

25

As the inquiry dragged on inconclusively, Harper's options grew fewer and fewer, and the Agency turned to ice around him by degrees. There was a certain patronizing smile that came over his colleagues when he met them in the halls, a solitude at lunches in the open cafeteria. Miss Lutkin could not hide her growing pleasure whenever the person he had her try to reach was unfortunately in a meeting or otherwise engaged. He had made no report of his investigation to the Deputy, but everyone in the building seemed to know that it was stalled.

He sent me a brief note thanking me for my advice, but he never returned to consult with me again. I would have been perfectly happy to give him guidance on the aggressive alternatives, but he sank deeper and deeper into passivity, hunched over his desk reading and rereading his files, trying to overcome his feeling that Birch was too innocent to take the blame.

There were gaps in the record. Inaudible portions of the transcripts. The important things eluded him, perhaps because they had at the time been the very things that in his trust he had never noticed, the pin-point lapses that might have warned him of Birch's potential.

Yes, Harper admitted that possibility now. Memory, like history, is really no more than a conceit about the past. It is created anew, an elaborate metaphor, to explain what appears now to be the case. The religious man looks back and sees revelation in what at the time may have only been a troubled dream. We shape our fathers in adulthood, as example or excuse. It is not that we lie to ourselves—though we do that too —because a lie must once have recognized the truth. Rather this is imagination in the void.

Finally the dread day came. The Deputy, who had not been in communication with Harper for weeks, called him to ask for a progress report.

"I'm afraid this will be brief," Harper said when he arrived for his appointment.

"Brief and decisive, I trust."

"Inconclusive," said Harper. "I'm at a dead end."

"But what about this Birch? Isn't he the real focus?"

"I just don't see his motive."

"Motive," said the Deputy. "Everyone has a motive. Use your imagination. What about that business in the brothel?"

"It doesn't track," said Harper. "After all, it was the Slav who took him there. Not us. We pushed him a little. That is true. But the balance of impropriety, for Birch's point of view, would be pretty even."

"He had another shock in that connection," said the Deputy, who was making no effort to hide his restlessness. Harper would have sworn the smoke from his pipe was darkening by the minute. "Did you plan to touch on that?"

"When the women were killed, you mean," said Harper. "There were some dicey moments, sure. But only because he has no tolerance for violence. He never seriously entertained the possibility that it was our doing."

"One of his weaknesses," said the Deputy.

"He trusted my word."

"And you his," said the Deputy.

"Within reasonable limits. There are none of the usual vulnerabilities. He was not rapacious. And as for ideology, Birch probably thought of Marx as that funny man in the movies with a shoe-polish mustache and a gift for insult."

"Politically unsophisticated," said the Deputy. "They are sometimes the easiest to sway."

"This is a man who loves to talk of his boyhood," said Harper. "He loves his country the way he loves his youth. Show me an apostate who had a happy childhood."

"It seems to me that you are straining, Richard," said the Deputy. "You really don't have much to show."

"I can't get near him. The Sisters think they have a case. I don't think they would mind very much prosecuting one of our assets. I get the surveillance take, of course, but it's been nothing. Watling reminded me that any direct contact I made with him could be seen as obstruction of justice."

"I see Birch is still code-word cleared. He might be pretty useful to the other side."

"He doesn't deal in anything more sensitive than medal citations. He's in personnel. The clearance is a formality."

"And musk to Kerzhentseff," said the Deputy.

"I don't say that Kerzhentseff wouldn't take what he could get. I thought I had covered the options pretty completely in the briefing paper. The question is whether Birch has any independent reason to go over to the other side."

"You covered the options," said the Deputy in the bored tone of a man of arbitrary power. "You've done a very good job of listing damned near everything. You even held on to the one possibility that you have a paper record clearly to refute— that Birch was somehow being run from this end."

"It is the alternative that would be most embarrassing to the Agency, I realize," said Harper. "And I understand that there is a very real burden of proof to be met. You are right that I have no evidence. I've looked at all the paper and it leads nowhere. I would be glad to put that in writing, formally, if you think that it would be useful."

"Don't try to corner me, Harper," said the Deputy. "I don't give a damn who gets red in the face. I thought I had made that clear to you at the outset. All I want is results. And if you try to use this cover-up line with me, I'll tell you right now that it will be your undoing. I want a disposition of this case. And I will not accept such things as your intuition or your charming loyalty to Birch as evidence. I am not entertaining guesses, any more than I am acting upon the rather widespread speculation that if there is a weak link in the case, it is very likely to be you."

"I didn't ask for the assignment," said Harper.

"It wouldn't make any difference if you had come begging me with tears in your eyes. You have the assignment, and the question is what you will do with it."

"I'm going to have to wait for the Sisters, I'm afraid. I can't get around that."

"You do what you have to," said the Deputy. And then he turned his back, the thick smoke from his pipe clouding the map of the world that hung above him.

When Harper returned to his own office, Miss Lutkin sensed his travail. She graced him with the knowing nod a schoolmarm gives the unruly child back from his talk with the principal. He asked her to hold his calls. Then he went straight for his couch, where he lay down and shook like a man with a fever. The telephone rang. Calls to return. This and that. Measurements of how close he was to the bottom now. Then the scrambler device gave off a rude little noise. Someone running a test, no doubt. Nobody was talking sensitive information to him these days. Nobody wanted to take the risk. The door opened.

He climbed upright to meet Miss Lutkin.

"It's the green phone," she said.

"It will wait."

"Mr. Watling is on the green phone."

"I hope you told him I was out."

"He says that fellow Birch is on the move again and that he thought you would want to know."

Harper turned away from her and punched into the line.

"Your friend like to take rides in the country?" Watling asked.

"What's going on?"

"The mark is somewhere on his way to East Jesus. There's been a contact, and we think he's going to make another drop. In the spirit of cooperation and mutual respect, I thought I should keep you fully and timely informed."

26

The rush hour was over as Birch set out on the parkway in the hazy morning sun. The wind whipped at his sleeve and rustled the clutter in the back seat, but it did not refresh him. He felt closed in, and the hot blast seemed to come straight off the burning engine into his face, a form of torture, a test of his will. The juice he had gluped on his way out of the house had come back up to his throat. He verged on a headache. The music on the radio was annoying. Up the dial and down again and nothing but punks and prayer.

Ahead he saw the big green sign:

CENTRAL INTELLIGENCE AGENCY
FEDERAL HIGHWAY RESEARCH STATION
NEXT RIGHT

He checked his mirror. It wasn't easy to tell whether anyone was following. The traffic traveled in clusters, so if somebody stayed behind you, it didn't necessarily mean anything. You could speed up or slow way down to see, but that would only draw attention to yourself and make them more careful. You could pull over to the side, but they would have a backup car ready to pick you up again when you eased back into the flow.

His instructions had said nothing about evasive action. He was simply to follow the assigned route and perform the specified actions. They would take care of the rest. He had to assume they knew what they were doing. He had to assume it, but that did not mean he had to believe it.

The first summons had come unexpectedly after all the years of silence and routine. He had given up hope that the chance would ever arise again, but then he had gotten the call that brought him back into the secret game. The instructions had been brief and specific. He had no chance to ask questions, then or since. He was operating on his own now, and while it flattered him, it also left him feeling vague and uncertain. It was hard now even to remember the caller's words with precision. He had been asleep when he felt Donna leaping out of bed to stifle the telephone's ringing. She had returned to say, not without some irritation, that it was for him. She did not know who it was, and the voice was unfamiliar to him too. The conversation, such as it was, had ended before he had awakened to what it really meant. By morning, he could only distinguish the event from a dream because of the note he had made and because Donna remembered the call too.

He had made the contact as ordered, despite his uncertainty. If it had been a dream, what harm could there be in playing it out? He wrote his simple message, left it at the prearranged spot, and walked on. He had been tempted to go back the next evening to see if the paper was still there behind the downspout. But that would have been a cardinal sin. He kept to his discipline and waited. A strange, unpleasant feeling of unreality crept over him as days and weeks passed with no further word.

Then on Saturday morning he had gone to do the shopping. The store was crowded. It always was. The ladies left their carts skew in the aisles and looked at you with an expression of violated privacy if you eased them to the side to make a way. When he reached the checkout, every register had a line. He got into what looked like the most promising. Then, behind him, someone spoke his name.

He turned but did not recognize the man, who was carrying

only two small jars of jam. The man told him to face ahead
and listen. Birch noticed the slight accent, believing it was
Russian, and the instructions for the next drop were given in
that tense and practiced way he had become familiar with overseas.

"But what do you want from me?" Birch whispered over his
shoulder.

"You shall provide what you can," said the man. Then he
disappeared, leaving the jam jars resting precariously on a
Family Circle rack.

The parkway merged into the interstate heading north. He
slowed down to take the long, deceptive curve, and the single
file parade behind him dutifully followed. Halfway across the
bridge over the wide Potomac Valley a sign welcomed the
traffic to Maryland and warned of the radar. Birch checked his
speed and edged it down a notch. He knew the reputation of
the Maryland cops, and now was not the time to get his first
speeding ticket. Not with what he was carrying in the glove
compartment.

It should have felt easier this time than it had in Tokyo.
Nobody was testing him against the polygraph. It was no trouble forgetting what it might become necessary to deny. The encounters had been so fleeting that the problem was trying to
remember them. In Tokyo he had learned his cover stories so
thoroughly that he could hardly distinguish them himself from
the truth. Now he wasn't sure what the truth really was. What
purpose was he serving? Who was his handler? Disembodied
voices spoke to him like the whispers of madness. Messages
were left for invisible communicants. He yearned to see
Harper again, Kerzhentseff, to sense the conflict as man against
man.

The traffic thinned as he got farther from the city. The highway was a long strip of white furrowing the wide, black, loamy
fields. Here and there a new corporate headquarters sprawled
on a manicured lawn. And as he drove on, even these disappeared. The land here was saved for farming. The roll of it
probably made it tricky to plow. In Cleanthe you had to plan
your contours on an elevation map because the rises and hol-

lows were too slight to be felt through the heavy grip of the tractor's treads. Here the land was hilly, and the farmhouses took the promontories. The earth looked rich, and Birch breathed in the sour, invigorating smell of manure and soil. A tractor trailed a wake of dust in the distance. They could use some rain, but the crops shimmered alive in the heat. He waved his hand at some boys fishing for bullheads and croppies from a culvert. One boy saw him and waved back. Birch was more relaxed now, and when his exit came, he slowed his car and took it easily. He barely noticed and easily dismissed the car a couple hundred yards behind him that made the turnoff too and after a longish pause at the stop sign followed him onto the two-lane state highway.

There were a few pickups on the road and one big combine leading a slow parade in the opposite direction. The fence along the ditch was good, some of it electrified. And the mailboxes on their weathered posts had little plastic cylinders with the names of newspapers from Baltimore and Washington. It was right for the people who lived here to want to keep the city at a distance. They allowed it to reach them only by dead drop.

The landmark for his next turn was a dirt racetrack, and it came up much more quickly than he had expected. They were racing some old wrecks on it, nothing fancy, just some kids having their innocent fun. On Sunday afternoons the old bleachers probably sagged under the ample butts of the farm wives and their men as the real, semiprofessional drivers showed their stuff. But today there were only a few people here and there seeing what the amateurs could do. The little refreshment stand wasn't doing much business. The sign on top said RACEWAY in letters leaning forward, apparently to suggest speed.

Birch pulled over onto the gravel shoulder and stopped to watch a filthy Ford roll across the finish line, exhaust cutoffs open wide for maximum noise. A sorry Dodge limped in behind. In Cleanthe they used to do their racing in the Crowleys' pasture where the ride was rough and the wheels kicked up a mixture of grass and mud and horse droppings so that it paid to hold the lead. He checked his watch. He was well ahead of schedule, so he turned off the key and stepped out from behind the wheel to stretch. His shirt was soaking wet, and gnats played on his matted brow. He left the car windows open and started across the road. It was good to be in farm country again where you didn't have to worry every time you left your car that it wouldn't be there when you came back.

Fishing for change, Birch trudged to the shack. It was tended by a busty girl who could not have been long out of school, but was already fading. She wore tight short shorts and a jersey halter that held her loosely. Her skin was tanned, but not in the hard, healthy way of a farm woman. Hers was sallow and dry from idle exposure.

"Beer?" she said.

She was not a pretty girl, and not very modest. Her nipples pointed insolently from behind the fabric. He looked away. They used to dress differently in the country. In summer the blouses and skirts would smooth out the finer details, making the curves and bumps only things to imagine. You had to strain to see the elastic under the white cotton.

"I asked you about beer, pal," she said in the strange, vowely accent of the border states.

"Excuse me," he said. "Yes. A beer. Mighty thirsty. I was thinking about something else."

"I bet you were," she said, leaning over, breasts suspended in the flimsy slings. "If you have something on your mind, you better get it out."

"What's the name of this town?" he asked.

"Dumpsville," she said, waiting for the foam to subside. "Nowheretown. The pits."

"It sort of reminds me of my hometown—back in Illinois. Except for the hills."

She paused a second and seemed to think he was making some kind of remark. "Sorry 'bout that," she finally said, letting it pass. "That's sixty-two."

"Here," he said and handed her a bill. She leaned over again for change, showing him everything. He was stricken with disgust.

"That's sixty-three, -four, -five. Seventy. Five. And one," she said. "By the way, my old man don't like wandering eyes."

"Your father?"

"My old man," she said. "There. In the Plymouth."

Birch could only see the big shaggy head, cigarette pasted in his lips, and a beefy, bronze arm dangling from the window.

"Thanks," he said.

The beer wasn't really cold, and it was more bitter than quenching. He leaned against the post of a hurricane fence at one end of the track, just out of earshot of a knot of youths idling around the Plymouth. One of them noticed him watching. But he just continued rolling a joint and lighting it with a long, conspicuous pull. When a greasy young man placed his hand flat against the tight denim ass of his girl and began to stroke, it was too much for Birch. He crumpled the wax cup, put it in the trash, and turned back to the road. It was then that he saw them.

"Hey! Get away from there!"

The two boys could not have heard him over the noise of the punks trying to impress each other with the sounds of full throttle. One boy stood at the window and the other rooted around inside.

"Hey!"

When they finally saw him running toward them, they scattered. But not far. They trotted off a distance, then joined forces again and watched him from just out of range, towheaded and surly.

"Oh Christ," he said.

The glove compartment had been opened and the contents strewn on the rubbery floor. He picked up a handful of papers and riffled through them. Maps, warranties, owner manual, scribbled notes. It wasn't there. He was frantic. What would they want with the documents? What do they know out here about such things? Then he spotted the envelope nosing out from under the seat. He slid it free and, trembling, checked to see that it was still sealed.

They had removed two of the four screws that held the radio. A third was halfway out. He found the missing parts and put them in his breast pocket. The envelope he folded neatly and slipped it into the pocket at his hip. Then he slid behind the wheel and stretched to lock the passenger door. The plastic upholstery was hot from the sun, and the wheel burned his hands. When the car started to roll, the boys leaped across the ditch to safety. But they fled no farther. And when he passed them, one of the boys—cute, freckled, pure Norman Rockwell —lifted his fist and slowly raised his middle finger.

Birch held the wheel tightly and stared out through the welling tears. He had lost his past utterly. He had lost everything. It was not simply remote in distance and time anymore.

They had changed him, removed him from the land, made him a lie. And now even the memory of Cleanthe seemed as false as he had become.

But it was only two miles more to New Market, and he had to get himself together. He had his orders: Park here. Stop here and here and here. Make the dead drop on the other end of town, an old Maxwell House coffee can on the side of the road just past the mile marker. They did not tell him why or to what end. They trusted him only to follow instructions. If something went wrong, it was up to him to find a story. They trusted him to be whoever it was necessary to be. And who he really was or had once been did not matter. He did not exist.

Birch breathed deeply and wiped the tears from the corners of his eyes with the back of his hand. The sweat came off his forehead into the gulleys beside his nose. But his heart was beating more slowly, his breath coming more steadily. He had his instructions. That was all he had.

He checked the mirror again when he reached the edge of the town. Somebody was back there, but it seemed a mile. The stores on the fringe were kind of seedy. Signs were white-washed on the windows of broken-down houses or nailed up against the mailposts. Closer to the center of town they became more respectable. The homes were restored and freshly painted. He drove past a house displaying stained-glass windows lighted dramatically from the inside. Next to it was an oppression of heavy wrought iron. He found the place where he was supposed to start and pulled in near it to park. He locked the doors, took a few steps, went back and checked them again. Then he patted his hip pocket and heard the reassuring crinkle of the envelope.

It was still early in the day, and the town was not crowded with shoppers. A few tourists were busy in a recreational vehicle regaled with stickers. Pairs of fine ladies from the city window-shopped here and there. A car passed. A man and a woman. Young, average type. Leisure suit and sundress. They could have been anybody.

Birch did not know why they wanted him to stop in the three places before moving to the site. Perhaps they wanted time to observe him, to make sure he had not been followed. Perhaps there were to be more instructions. It didn't do any good to guess. He had to take it as it came.

"Don't be shy," said a tiny voice behind the brown-and-white gingerbread door of the Come Into My Pantry as he peeked inside. "I won't eat you up."

She sat in an intricately carved rocker, its heavy wooden limbs and claws making her seem all the more frail. Her hair was blue, too blue, and her lined face was powdered white as a clown's.

"Something for the Mrs.?" she asked.

"Yes," said Birch. "I thought I might just browse."

"Pie coolers, pots, pickle jars, paring knives," she said. "We have everything there was. My husband and me, that is. He's dead."

"I'm sorry," said Birch.

"No sorrier than I am," she said. "He just gave out like an old car."

Birch moved away from her.

"It's all arranged along the walls," she said. "Andirons to the Zenith over there. That old radio still works. You can turn it on if you want to be sure."

Birch cast his eyes over the implements. Iron, pewter, rusted steel. If there were a method of organization, it eluded him. The heavy, old things lay on tables and shelves in a clutter of remembrance. A grain grinder like the one they used at home for flour, an iron skillet. It could have been their summer kitchen on the farm, except for the disorderliness and the fear.

"It's all original," said the old woman, "just like it used to be."

"Does this still work?" Birch asked, holding up a sooty kerosene lamp like the one his grandfather had kicked over in the privy out back when Birch was a boy. The privy had burned. His grandfather had said that he tried to piss it out but that he just wasn't the man he used to be.

"Smokes like a politician," she said, cackling.

Birch felt the strange old woman's eyes following him around the room. He did not know how long they expected him to stay or whether this unlikely woman bore him some message. He forced himself to pick up an item here and there. The fittings on the moving gadgets seemed perfectly good, if a little rusty and stiff. The metal would shine up. A bent ladle, a speckled colander, an eggbeater like the one his mother used for making frosting. He could almost taste the galvanic metal beneath the sweet the way he had when she gave him the utensil to lick clean. But the taste went sour. He put the eggbeater down. If the old woman was the one, if she had words for him, let her speak them now. He turned and hurried to the door.

"A little baby's spoon?" said the harridan's voice at his back. "A catch basin? A kettle? It's all very clean."

When he reached the street again he realized how cool it had been in the dank old house. Their summer kitchen had been like that. Cool and musty like a cellar. The children would go there in the afternoon before mother stoked up the stoves. They would play games in the shadows where the only dangers were imagined. He cleared his throat and wiped the water once again from his eyes before crossing the street to his next stop.

The Morticed Board. A sign in the door window said OPEN, with a silly smile drawn in the O. He stepped inside. Instead of mustiness, this big room crammed with breakfronts, commodes, old wooden iceboxes, and lowboys smelled of sweet pine and varnish. The proprietor was a tall man about Birch's age, in jeans and a Perrier T-shirt. The tattoo on his forearm might have been an anchor or a cross. It was blue as a bruise and shapeless with years. The plume from his cigarette teased his eyelids as he bent over a wooden chest polishing the brass fittings.

"You came to the right place," he said.

And Birch, in case it was some code, replied, "It comes highly recommended."

"Farm furniture. Best there is. I pick it out myself at the sheriffs' auctions. Takes a sharp eye for value—what can be restored, what you might as well forget. Sometimes the folks have slapped paint on, jammed the piece in a wet corner to rot. But look here," he said, pulling a drawer out of a bureau. "See that craftsmanship. All done by hand. Made to last. Stripped this baby down myself. Gave it a nice fresh polyurethane finish. Good for another hundred years. I'll let it go for four hundred."

"My parents had a piece like that," said Birch.

"They get it from me?" said the proprietor.

"What? Oh. I'm sorry. No. They bought it new, or the grandfolks did."

The man creased his brow and looked Birch over carefully. Then he stubbed out his cigarette in a Campari ashtray and put the polishing rag down next to the cash register.

"Now I ask you whether it makes any difference whether a piece has a few new slats or a rebuilt leg? A man on horseback isn't going to know the difference. And let me just say this: My price is right."

"I'm not really in the market for anything large," said Birch.

"Three seventy-five and I'll deliver it myself."

"I don't think so."

The proprietor slapped his hand smartly against the shining finish.

"I can see I had you figured wrong. What did you say your name was?"

"Anderson," said Birch.

"Come along with me, Andy," said the proprietor. "I can tell that it is real quality you are after. I keep the nicer pieces back here."

Birch followed him out of the showroom through a corridor filled with half-finished tables and stools and church pews stacked one on top of the other. The proprietor led on into his living quarters, which were cramped with furniture, low-ceilinged, and dark. Birch followed because he did not know what to expect, where the critical contact might be made, who it would be. He went where circumstance took him, like God's own fool.

"Now this is one you'll appreciate," said the proprietor, running his fingers over the delicately joined boards of a corner cupboard that stood at the farthest end of the dining room. "It's original. The stain's original, too. They used blood as a base. It's what gives the finish its depth."

Birch moved closer and pretended to inspect the details of the cupboard. He opened the door and looked at the joints, how the doors were hinged, because it seemed the plausible thing to do.

"I'll be frank with you," said the proprietor. "One of the legs was badly cracked and I had to tool a new one. Hard as hell. See, they're solid oak spheres. Getting the stain right is a bitch. Try to go out and buy a can of blood sometime and you'll see what I mean. But here. See if you can tell me which one is the new one."

Birch kneeled down on one side of the piece and then the other.

"The one on the left," he guessed. "The rear one."

"Still," hurried the proprietor, "you got to admit it is a hell of a close match."

"You did a nice job."

"Because of the gimpy leg, I'll give you a super price."

"I don't think so."

"Look, Andy," said the proprietor. "Take my card. Have a little talk with the wife. Give me a call in the morning with your decision. I'll hold the pieces you're interested in."

"That's very kind of you. But don't go to the trouble. Is this the way out?"

As Birch retreated from room to room, the proprietor kept up a barrage of offers and discounts, sales points, little stories about the history of this item and that. Birch fled through the showroom door without another word.

His last stop was a bookstore fifty yards down the street. It was in a huge Victorian house whose spires were dingy with flaking paint and tilted with age. Its windows were grimy. The hand-painted sign hung askew. Birch paused at the door and noticed, reflected in its window, two men in business suits sipping coffee at an outdoor table of the café across the street.

He pushed open the door quickly to get away from them and was startled by the piercing jingle of the chimes. He looked up at the bells, still moving but silently now, and then he let the door close, setting them off again.

"Hello," he said. There was no answer in the dusty hallway. He felt like an intruder. He stepped forward as lightly as he could. The floorboards creaked and groaned. On either side of the center hallway the rooms sagged under the weight of thousands of faded volumes. Every surface—tabletops, shelves, windowledges—was stacked with them. Leather, buckram, paper. They spilled off into piles on the floor.

"Hello," he said again. And from the far end of the hall he heard a fragile cough. "I thought I'd just have a look around, if that's all right."

A tiny bald head, bespectacled, with elfin ears and a tiny nose appeared around the corner.

"Fine, fine," said the head.

Birch went to the shelves and skimmed all the old, obscure titles. As much time as he had spent in libraries during idle army hours, he did not recognize any of them. There were whole sets of the collected works of dead men with names like Severson and Puce. There were morocco-bound histories of forgotten places, biographies of people whom words could not preserve. He ran his fingers over the ribbed spines with their gold lettering. Fancy caskets.

A doorway led toward the rear. He took it. The house was a warren of little rooms, all of them dark and creaking, heavy with the yellow smell of old paper and mildew. He finally found himself in a cul-de-sac lighted by a single bulb through a water-spotted shade. There were children's books there, ancient stories of rabbits and ducks, illustrated with silly, literal drawings of the sort that were all right in the days before the animals began to move and speak on the Saturday morning cartoon shows on TV. Little boys in sailor suits, looking innocent and surprised, populated this book. Girls in pinafores danced through that one. They were like the books he had leafed through as a child, handed down by parents and grandparents, read to him sleepily at night to stoke his dreams. Birch was fascinated by these adventures, stories of journeys under the sea, flying teacups, caves. He barely raised his eyes from the pages when the chime gave its double warning.

The other man was in the doorway before Birch registered his presence. It wasn't one of the people from across the street.

"You will make the exchange now," the man said.

He was big and shapeless, his shirt darkened under the arms by exertion and the heat. He wiped his forehead with a crumpled, dirty handkerchief.

"Some weather," said Birch.

"Your instructions are changed," said the man. "I will take delivery, make payment."

"I'm sorry," said Birch. "I don't understand."

"You will not go to the drop site."

"What drop site?" Birch asked. He had been given no other means of identification.

"Milepost forty-three," said the man. "Please hurry."

"Is there danger?"

"There is always danger," said the man.

"Shouldn't you pick something out? Here, one for your children."

"Silly dreams," said the man. "The document."

Birch reached into his pocket and pulled out the rumpled envelope.

"I hope this will please them," he said.

"We will be in contact," said the man, and he put a roll of old anonymous bills in Birch's hand. Then he turned and took three steps before the others emerged from the doorway, pistols drawn, shouting, "Hold it right there!"

And though he was not sure why, Birch clung to the little chapbook about a boy and a bear, the one remembered tale among the dead, and he felt an enormous relief.

27

"Sometimes you've got to make the draw. It's part of the job description."

The white walls of Watling's office were bare except for the photos of the Directors, old and new, the latest President, and a framed child's stick figure of a man with an enormous hat. The drawing was inscribed, "Be brave dady."

"Did you make the decision?" asked Harper.

"It was made on site," said Watling. "They didn't expect a hands-on contact. The plan was to check out the drop before it was cleared. We figured we'd either let the shit pass or stop it if it was too good to let go. But we didn't have a chance. The pass was in the dark. My guys decided they had to make a move."

"They lost their nerve."

"We don't take risks," said Watling. Then, for good measure, he added, "We don't get doubled either."

"What did he turn out to be handing over?"

"That's the problem," said Watling, dropping his guard. "I thought you said this guy was code word cleared."

"He has them all."

"Then what's he doing driving halfway across the state, bob-

bing and weaving our asses off, ducking into a place that's like
something out of those rainy paperbacks my wife is always
reading, and then passing this junk?"

He tossed a Xerox on the desk. It looked to be a duty roster.
Classified secret, of course, but only because somebody must
have liked the color of the stamp.

"Maybe there's something under it," said Harper, deadpan.

"You train this guy in secret writing too? Jesus. You probably gave him the key to the executive can at Langley. Hell of
an instinct for people you guys have over there. No, there's no
more than meets the eye. It's crap. Classified crap."

"In Tokyo we started him out passing small stuff. Duty roster was part of it. That's the way it's done, you see. It is something I would have been glad to point out, along with the fact
that Birch did not really have any access to significant information in his current job."

"Maybe we didn't exactly trust you," said Watling.

"It doesn't say much for your instinct for character either, I
guess," said Harper. "I don't suppose Justice's enthusiasm for
spy trials had anything to do with your hair trigger on this."

"We get a little sick of the memos you guys keep sending
over complaining about our lack of prosecutorial vigor. You
probably wrote the damn things yourself, for all I know."

"Not me," said Harper. "I'm your biggest defender."

" 'In the past fiscal year,' " Watling intoned, " 'the number
of Soviet and Eastern Bloc personnel entering the United
States on student, cultural exchange, and other special visas
has increased by one hundred and fifteen percent. In the same
period there has been only one espionage indictment. Two
diplomatic personnel have been expelled as persona non grata.
This represents a net decrease of thirty-three percent.' You're
keeping score, for Christsake. And the message is that the
blood is on our hands when the Red Army marches in jack-
boots up Pennsylvania Avenue. Then, when we do make a
move, State pulls out its pince-nez and reads us a long lesson
in détente. They've already been complaining about the bust. It
seems the courier travels on a diplomatic passport. No matter

that we have a file on him this thick. No matter that he's KGB down to his socks. You'd think we busted Gromyko."

"Can you make a case against Birch and the courier?"

"Out of a duty roster? Nobody's going to buy that. The defense would just get it declassified. Hell, they could probably get it through the Freedom of Information Act. We'd like to expel the other guy, but I bet even Ivan beats the rap."

"Nice work," said Harper.

"Your friend isn't helping very much either. Malloy has been questioning him. Birch is an odd one. He doesn't want a lawyer. He acts like he's cooperating. But he isn't giving us beans."

"He's used to going it alone."

"Meaning you trained him well," said Watling. "I guess I have to give you credit. You're the one who found this mope, gave him access. And I'm the one who's sweating."

"Can I see him?"

"They've got him over at the field office. We might as well take a walk over. It isn't going to do any harm."

The sky was gauzy with haze and the pavement was overheated. Harper took off his coat and hung it over his shoulder. Watling kept his on, probably to cover the sidearm. It was the kind of day when women forgot about modesty and wore as little as they could. But it was hard to work up much enthusiasm about that in the heat. An exhaust fan blew the greasy smells of a hamburger joint into their faces. A junk shop on the corner blared rock music into the street. But nobody was dancing. The desolate men sat on doorsteps in the shade waiting for their connection, the night.

Being with Watling, Harper had no trouble with security at the field-office door. They rode the sweaty elevator to their floor. It was cramped with others, each trying vainly not to touch. When they entered the intelligence unit squad room it was as if they had surprised a bunch of conventioneers in a raid. The voices went silent. Laughter choked off. The air conditioner ticked and strained. The agents stood in a stony tableau, shirtsleeved and holstered, ready for punishment.

"This is Richard Harper of the Agency," said Watling. "Some of you may know him already." Torchiana scowled. "Has there been any progress?"

"Nothing, sir," said one of the agents. "He's a tough one. Best we can get out of him is that business about the Agency tie."

"Won't do," said Watling. "They've got their shit squared away."

"Yes, sir."

"Where is he?"

"Interrogation room two."

"That have a two-way?"

"Yes, sir."

"Well, you seem to have done one thing right," said Watling. "Let's see if we can't build on that. Has anybody talked to Justice?"

"I have," said Torchiana.

"And . . . ?"

"They weren't impressed. No way we can hold him the night."

"You disagree?" said Watling.

"No, sir," said Torchiana. "Absolutely not. We're coming up dry. But I do know this. That dude in there is a bad one. Lucky maybe. But bad."

"Or smart."

"That too."

"Smarter than you were maybe."

"Maybe."

"You thought your eye was fast enough to follow his hands. You lifted the shell and expected to find the pea. But what was there?"

"Nothing."

"He faked you, didn't he? You all left your jockstraps back on the ground in New Market."

"Yes, sir."

Watling led Harper down a long hallway broken on either side by numbered wooden doors.

"I thought you stood behind your men," said Harper.

"I do," said Watling, "except when I'm standing in front of them."

He opened one of the doors and waved Harper into the shadows. One wall of the little closet was lighted up. The two-way mirror had the effect of smoked glass, making Birch and his inquisitor seem dusky and remote.

"You'll have to keep your voice down," said Watling. "We listen in on mikes, and the wall is soundproofed, but if you howl they can hear it through the mirror."

Watling found the switch next to the pane and flipped it. The voices came on, and it reminded Harper of that last night of the operation in Tokyo when all he had to worry about was whether Birch would get away with his lies.

". . . and had you checked out," Malloy was saying, reciting the words flatly, like a priest at mass. "The CIA has made no contact with you. We have made no contact. The records are very, very complete. Now, do you want to try again?" It had reached that stage already. The only technique they had left was repetition and fatigue. Birch could handle it easily.

Still, Harper could see that Birch was not relaxed. He held his forearms flat and still on the arms of the chair the way he had been taught to do when he was wired. His body was steady and his breath shallow and regular. But there was a tension that cut into his face, lined and aged it, gave him away to anyone with eyes to notice.

"Do you have a stress machine working?" Harper asked.

"Don't believe in them," said Watling. "Damn things are as likely to give a bad reading on the agents doing the questioning as on the subject."

Malloy paced back and forth in front of Birch, who followed him with tired eyes. Harper had never seen him this way before, mazy and beaten. He had held up against so much, but now he seemed broken, and Harper did not believe that their questions had been enough to shatter him.

"Let's start over at the beginning," said Malloy. "When did you receive the alleged phone call?"

"Which call is that?" Harper whispered.

"Wait," said Watling. "You'll hear."

Birch paused, as he had been taught, then answered. "It was two days before I made the first drop at the Hay-Adams. It was late in the night and I was already asleep."

"Did you recognize the voice?"

"No," said Birch. "It was a man's voice. But it was kind of strange, metallic-like. I kind of thought it might have been electronically changed."

"But you knew it was bona fide."

"Yes."

"You didn't become suspicious at an unfamiliar, probably altered voice telling you to get cozy with the Russians again."

"It was the CIA. I was sure of that. They needed me again. I didn't know why, don't even know now. But I knew it was them."

"How were you so sure?"

"I knew, that's all. There was a way I could tell. But I can't explain it to you. I'm sorry, but I've promised never to reveal it."

"On your honor," Malloy sneered.

"That's right," said Birch. "There are certain things that I have been told not to talk about."

"You had better start reassessing your position on that," said Malloy.

"If you are who you say you are, this will all work itself out. You'll see," said Birch.

"You want to look at the badge again? It says United States of America on it. We're all one government, you know."

"I'm sorry," said Birch.

Malloy walked behind him and when he was out of sight stretched his neck stiffly and pulled the knot of his tie farther down. Birch wetted his lips with his tongue and closed his eyes. The pain looked real to Harper.

"This strange voice," Malloy said, "what did it say?"

"Like I told you, I can't remember every word. I was kind of groggy. But I was told to go to the hotel at a particular time and put a note behind a downspout in the alley. I was to report my current assignment. It was for the man I went up against in Tokyo, Kerzhentseff."

"And you did this."

"Yes. But, like before, I was really trying to fool him."

"A real patriot," said Malloy. Harper knew he was using the wrong approach. He was making the usual mistake. He had read Birch's file and was underestimating him. Malloy was playing to Birch's strength. "When was the next contact?"

"Just a few days ago. I was approached in a grocery store, given orders for the next drop. The man left some jars behind. I thought of taking them. For fingerprints, you know. But I didn't want to risk it in case someone else was watching me."

Clever, clever Birch. Making himself vulnerable, suggesting the next question, playing the inquisitor like a harp. He was tired and broken, but he was not defeated.

"You were afraid maybe that we were watching," said Malloy. "You were afraid you would be caught."

"I figured the CIA might be there," said Birch slowly. "But I was worried that Kerzhentseff might have other eyes. If the CIA wanted the jars, I figured they would pick them up later. I didn't want to look suspicious to Kerzhentseff."

"A regular junior G-man," said Malloy. "What were your orders? What did the Russians want from you?"

"Whatever I could provide," said Birch.

"They didn't specify?"

"No," said Birch, a nice note of hesitation in his voice. "They knew my work. From before."

"Didn't it seem strange to you that they were pushing you without even knowing what you could deliver."

"Maybe a little," said Birch. "Maybe just a little. But we were in America now. I figured it was harder for them here. They couldn't get around as easily. It was more dangerous."

"Yes it was," said Malloy, the pride justified only as bluff. It was transparent, and Harper was sure Birch saw that too. "How did you decide what to give them? Did the voice on the phone authorize it?"

"I never heard from him again."

"So you just grabbed up a handful of documents and off you went."

"I tried to think it out," said Birch. "I had to make my own decisions. What would the CIA want me to do? So I found a document, the roster, just a bunch of names. It was classified. It proved that I was working where I said I was, and in Tokyo we had started out with this kind of thing. I didn't think the roster would do any harm."

"You took it upon yourself to decide what was harmful and what was not."

"I didn't have any choice. I was on my own," said Birch.

"Isn't it true that the duty roster was the only classified document you routinely had access to?"

"I could have gotten others," said Birch.

"You gave them the best that you had, the most damaging piece of paper you could lay your hands on."

"I could have come up with others. I gave them the least I could."

"But the roster was the only thing that routinely came your way."

"I don't see the difference," said Birch.

Harper had to admire Malloy's persistence. He was working like a prosecutor, trying to find the ambiguous points, trying to turn them against the witness: the unnamed, mysterious voice on the phone, the fact that Birch had given everything he had, the foolishness of an enlisted man thinking he was supposed to run his own double-agent operation.

"I didn't have routine access to other things," Birch admitted. "That's true. But I was cleared."

"You knew how to get more highly sensitive material if you wanted to?"

"Sure," said Birch. "Yes, I did."

"You had figured out how to keep the spigot open. So later on you could deliver something better if the price was right."

"The price had nothing to do with it."

Malloy held out a roll of bills between his fingers.

"How much do you suppose is in here?" he asked.

"I have no idea."

"You didn't even know what the salary was going to be. No wonder you gave them so little. Just a taste, a sample to begin with. Then you could begin the bargaining."

"I never kept the money," said Birch. "I always turned it over to Mr. Harper. You ask him."

His name on Birch's lips was jarring to Harper. There was something awful about Birch's trust that he would corroborate, lend support, when all the while he was surreptitiously a part of the inquisition. It was faithless and raw. Watling smiled beatifically.

"You were going to turn the money over?" said Malloy.

"Yes."

"Who to?" Malloy shouted. "The voice on the phone? Your favorite charity? Internal Revenue?"

"They would let me know the procedure. Somehow, when the time was right, they would let me know." His faith was genuinely touching, whether or not it was genuine.

Malloy was getting nowhere, and Harper knew that, prosecution or not, more would have to be shaken out of Birch. There were memoranda to be filed, testimony for the committees on the Hill. Whatever the punctilios of the criminal law required, the Agency did not believe in pursuing only those conclusions which could be established beyond a reasonable doubt. All doubts were reasonable. Some were mandatory.

"Why don't you let me try?" Harper whispered.

"Try what?" said Watling.

"I trained him. I may be able to take him down."

"We'll give it a shot," said Watling, and he showed Harper the way to the interrogation room.

The lights inside were painfully bright. Birch did not acknowledge him in the slightest when he walked through the door.

"Hello, Jerry," said Harper.

"Hello," said Birch, flatly.

"This guy says he's a friend of yours," said Watling.

"Why is it that I just knew I'd see you here, Harper?" said Malloy.

"Cop's instincts," said Harper.

"I guess. Your buddy here claims he was following Agency orders."

"So I understand," said Harper. "Do you mind if I ask a few questions?"

"Go ahead."

"Jerry," he said, "you don't have to pretend. As you know, they have my name. They know about the Tokyo operation. They know who I am. They have read the file. We gave it to them. It is all on the level. There's no trick."

Harper handed him a handkerchief to dry the sweat from his face. Birch seemed more than confused. He was close to panic, looking from face to face in fear, or a convincing imitation of it.

"Something has happened that as yet does not make very much sense," Harper said softly. "With your help, perhaps we can straighten it out. First, you must understand that you can be perfectly frank. You need not hide anything on the Agency's account. We are all in this together."

"All except you, Birch," said Malloy, the bad cop to Harper's good. "You are alone in a world of hurt."

"I'm sure it is only some kind of misunderstanding," said Harper. "You perhaps thought you were doing the right thing."

"Yes," said Birch. "You gave me a medal."

"And you deserved it," said Harper. "Deserved more." The man had become before his eyes as simple as a boy. It was pathetic, or else it was genius. "Now let's begin with the telephone call. Whose voice was it?"

"It's a funny question," said Birch.

"You see me laughing, don't you?" said Malloy. Harper gave him a silencing look.

"Feel free to answer, Jerry," said Harper. "It is very important. We have to see if we can find out how this thing got started."

"I didn't recognize the voice," he said, carefully. His eyes seemed to plead with Harper for instructions, for some subliminal sign.

"You said that you knew it came from the Agency, that it was legitimate," said Harper.

"Yes," said Birch, and then repeated it. "Yes, I did."

"If you didn't recognize the voice, how could you be sure?"

"I didn't know why you would want to alter your voice. But I figured there must be some reason."

"Did you think it was me?"

"Yes," said Birch, and it was as if a weight of years had been lifted from him. His whole body relaxed.

"Why did you think so, Jerry?" Harper asked. "Did the man identify himself by name?"

"It was the word you used," said Birch.

"I want to be straight with you, Jerry," said Harper. "I did not make the call. But if someone did and used my name, we will want to find out who it was."

Harper hated the smugness he saw on Watling's face.

"The man didn't use your name," said Birch.

"Then what made you think . . ."

"He used the word, Mr. Harper."

"The word."

"The word we used in Tokyo, just you and me. You said no one, absoutely no one else knew it. Otherwise, maybe I would have wondered. But when the man on the telephone used the word, I just knew it was you. And I did what you said."

"What's the goddamn word he's talking about, Harper?" Malloy asked.

They said it very nearly in unison, Harper and Birch. The word was *convergence.*

28

When she saw his dazed eyes and smelled the sour-mash sweat on him, Donna Birch wasn't sure she could handle him, and she was scared. He lurched through the apartment, banging knee against end table but showing no signs that he felt it. The little figurine samurai and geisha wobbled. His hands lunged to save them, sent them shattering to the floor.

"Shit," he said.

"It's all right, Jerry," she said, taking his arm. He wrenched it away and stumbled into the kitchen. After she swept up the fragile, precious shards—doing it quickly so as not to indulge in their loss—she joined him there. He stood at the counter pouring himself more. Some of it was going into the glass.

"I don't think you need that," she said.

"Need something," he mumbled. "Get straight."

The whiskey troubled his face, puffing it up and twisting his lips the way a stroke would. She knew the way drinking could destroy a man. She had seen it happen to the husbands of too many of her friends. It stole up on them gradually. A little extra every night. Then a weekend toot. You hated to nag about it. A man had to be a man. And a wife should never be a scold. But pretty soon the whiskey was the wife, and the woman, well, she was just simply shut out of his world.

"Maybe if you lay down it would help," she said.

"Damn spinning," he said. "Can't seem to get things straight."

She wondered what Jerry had lost. If it was money, they didn't have enough to begin with to be upset about. His job was secure. Hadn't he gotten all his promotions early? The only thing of real value they had was what they shared, and it frightened her to consider that he might think he was losing her.

"Come with me," she said. "We'll go to bed."

It was still early. She had to grant him at least that much. Wherever he had been since he went off mysteriously in the morning, he had had the presence of mind to come away before closing time. He had somehow found his way home. That instinct, at least, was still functioning. She cringed to think what might have happened to him behind the wheel in such a state.

She held him across the shoulders and tried to guide him away from the walls. He stopped, tipped the last fiery drop into his throat, then came along. He teetered, and she held him tight. She had to handle him, scared or not. There was nobody else. If he had fallen, she would not have been able to lift him up. But he did not fall. He stumbled and cursed and stalled. But he did not fall.

"What is it, Jerry?" she asked when she had maneuvered him to the edge of the bed and sat him down on the intricate chain stitch of the quilt she had made to celebrate a better time. He buried his face in his hands and held it there as if two hands could steady the world.

"Don't know," he said.

"Don't know what?" she said.

"The trouble. Don't know."

She put her hand on his shoulder and rubbed there. Perhaps he could feel it just a little, feel the affection, even if she could not touch the source of the pain.

"Is it something at work?" she asked.

"Caught me," he said. "I let them down. Caught me. Under arrest. Me. Trying to be careful. I don't know."

"Did you have trouble in the car, Jerry? Did you get stopped by the police? It will be all right. Don't worry about it tonight. We'll find a lawyer. Let's see your wallet. Let's see what the ticket says."

She swung his legs up onto the bed, deadweight. When he was flat, she gave a heave and turned him to his side so she could fish the wallet from his pocket. Awkwardly, she got it between her fingers and released it from where it had snagged on the buttonhole and seam.

"Take it," he said. "Hold on to it. It tells who I am."

There was money in it, not much—five, ten, eleven, twelve dollars—but at least it showed that he had stopped drinking before it ran out. That was something, at least. She looked through the compartments for the traffic ticket that would tell her how bad it had been. But she found his license instead, the one with the dreadful photograph that made his face look lumpy and sullen. For once the image fit him, all puffed up and loose from the whiskey as he lay flat against the pillow.

"You're all right," she said. "You didn't get arrested. You must have imagined it."

But by then he was gone, comatose, beyond recall. She did not even try to get him out of his clothes or under the covers. She turned him over onto his belly in case he became sick, removed his shoes, loosened his collar, and spread over him an afghan she retrieved from the mothballs. Then she kissed his forehead where it was lined with his secret care. The alcohol on his breath and the sharp fumes of naphtha mingled. They were strong and heady. She breathed them in, as if the vapors could raise an oracle to his demon. But all they did was awaken her, smelling salts, sharpening her fears.

It would do no good to try to lie with him. The sun was barely down, and she was alert and beset. She left the table lamp switched to the dimmest setting, turned off the others and went to her place in the living room.

"Little boys," her friend had told her. "They are always little boys getting into mischief. Don't blame yourself—or them.

They do what they do just because they know they aren't supposed to."

Donna picked up the telephone to call her. The woman was so earthy and wise. But she couldn't dial the number. She was too embarrassed for him.

This was more than just a silly fling. He had never been a little boy, not even when he was a child. You never caught him getting into fights or playing chicken on the railroad trestle. He was steady, always had been. But now he had such moods. Too high. Too low. He had been elated after the phone call that came in the middle of the night. And now he was haunted.

Donna retreated from what she did not know. Finally, it wore her, drove her down. She was almost asleep on the couch when she was startled by a knock on the door.

"Who is it?"

She looked through the peephole at a well-dressed, middle-aged man in a suit and tie, his face distorted by the lens.

"Richard Harper," the voice said through the bolted door. "I'm a colleague of your husband's. Is he there?"

"Come in," she said. "He's asleep, but you can come in a minute if you want."

"Thank you very much, Mrs. Birch. I'm not surprised that he is tired." The man smiled patronizingly as he came through the doorway. She did not like him at all. "It was quite a day for him. Quite a trying day."

She showed him to the armchair where Jerry usually sat and offered him a cup of coffee.

"It's instant," she said. "It won't take but a minute."

"No thank you," he said, still smiling. "I won't be staying long."

There was something in his tone that upset her. He spoke in the empty, efficient voice of a doctor getting ready to break the bad news. But she tried not to show her feelings. He was a guest in her home, after all.

"Was there something I could do for you?" she asked. "Some message for Jerry? I don't think he is going to be waking up until morning."

"I wouldn't disturb him, Mrs. Birch," he said. "I hope you'll forgive me for saying so, but I feel as if I know you, from all the things your husband has told me."

"He never mentioned you," she said.

"I suppose not," he said, and she thought she caught him averting his eyes just a little. "He's going to need you now especially. This isn't going to be easy for him."

"What isn't?" she asked. "What's this all about?"

Harper cleared his throat and shifted nervously in his chair. The ingratiating smile was gone. At least she had made that much progress.

"I assumed that he had told you."

"He came home drunk tonight," she said. "If you know him so well, maybe you could tell me why. He could barely say his own name."

"I'm sorry."

"There's no need for that. He'll feel pretty badly tomorrow morning. But he'll survive."

"Of course."

"I don't know who you are or what kind of trouble Jerry's in," she said. "He had some crazy idea about being arrested."

"Yes," said Harper. "There's been a problem. Perhaps it would be best if I let him tell you about it." He was rising off the chair when she waved him back down again.

"You can't come in here and scare me this way and then just go off. It isn't right."

"Your husband worked with me in Tokyo," Harper said.

"I never met you, did I?"

"I don't think so, Mrs. Birch."

"I would have remembered," she said. "I'm good at faces."

"Just like your husband," he said. It grated on her. She had known so many condescending ones like this man, polite and snobbish, officers' wives. She remembered people because she took an interest in them. Jerry's memory was nothing like hers. He was intense but selective. He remembered painful things, things she never even noticed. The only way you could confuse the two of them was if you didn't really care at all.

"What kind of work did you do together, Mr. . . . What was your name again?"

"Harper," he said. "Richard Harper. It is a little difficult to explain."

"I'll concentrate," she said. "Maybe if I try real hard I'll be able to understand."

"I didn't mean it that way, Mrs. Birch. You see, our work was very secret," he said.

"Everything Jerry does at work is secret from me. That's the way the Army is. I accept that."

"I'm not with the Army. I'm with the Central Intelligence Agency."

"Well," she said, refusing to give him the satisfaction of seeing her shock, "it's all the same, isn't it?"

"It's refreshing to hear someone say so," he said. She had not meant to flatter.

"The job your husband did," he went on, "it was very important . . . and very sensitive. I can't really go into the details."

"Jerry keeps it all inside too," she said.

"Good," he said. "But recently there's been a problem. Your husband has been seen keeping the wrong company."

She lowered her eyes because she shared the shame. Jerry had always been so careful about his friends. But now he had fallen in with some bad ones, and they had gotten him drunk and into trouble.

"I'm sure it won't happen again," she said. "He's going to feel terrible about it. He really isn't like this, getting drunk and all. I know him well enough to know this will cure him."

"I'm afraid it's worse than that," he said. "He's been associating with foreign agents."

"Why would he do that?"

"I don't know, Mrs. Birch. I thought perhaps you might be able to explain."

"Is it that corporation again?"

"Corporation?"

"The one Mr. Nowicki worked for in Tokyo."

"In a sense it is."

"I can tell you why he associated with them," she said. "They treated him with respect. I always hoped they would offer him a full-time job some day. Paris, Rome. I'd like to go to those places. But they never did. Unless that was what the man called about the other night."

"Did Jerry tell you about the call?"

"I answered the phone, but I didn't pay any attention. Jerry seemed excited about it, though. He said the call came from a crank, but he was so thrilled that I knew better."

"Did the man on the phone have an accent?"

"You mean like Mr. Nowicki? I don't think so."

"Were there other calls?"

"None that I know about."

"Did he ever mention having heard from me?"

"No," she said. The vanity of this man was disgusting. "He never mentioned you at all. Ever."

She did not know what this man, Harper, was driving at. But it had a bad feeling, whatever it was. He looked at her with an expression that could have been sympathy. Or it could have been relief. Either way, she loathed it.

"You've gotten him into some kind of bad trouble, haven't you," she said.

Then, for the first time, she saw something genuine and touching come into his eyes. He held her gaze for a moment, then looked away. It gave her a chill, what he said next. The very fact that he had in this instant become a person for her, somebody she could imagine caring enough about to remember, made the trouble seem all the larger. It was bigger than this person or that. Jerry was in great danger, and she did not know why. She had accused this man of creating it, and all he said was:

"One way or another, I think I have."

29

Before returning to Langley, Harper stopped at a telephone booth outside a drugstore near the Birches' apartment. He would not be able to see Fran tonight. There were too many things he had to do, for his own self-protection. But he wanted to talk to her, to explain what had happened, the call from his wife, to apologize.

He left the door of the booth open to catch the breeze and dissipate the acrid smell. The vandals' art darkened the plastic windows: Magic Marker equations of love, telephone numbers and names, promises of what used to be called "unnatural acts" before the discovery of the superior ethic of dreams. There was no answer at Fran's, so Harper dialed his own number just in case she had forgiven him and let herself in again. He let it ring ten times, counting them, before he gave up.

All he had accomplished was to give them another tantalizing item to report. "Subject abruptly pulled to curb, proceeded to pay phone, then dialed two numbers. Nature of communication: unknown." Unlocking the door of his car, he scanned the street for signs of surveillance. Nothing. The Sisters were very good.

He noted the time and the number of rings. If they were

watching him, he would preempt them by explaining the event in his report. And if they were sitting on his telephone line at home, the details would provide some corroboration.

This was not paranoia, Harper told himself. It was not the devil cult of possibility. This was purest realism. Birch had named his name. He was no longer the investigator; he was the prime suspect.

The secretaries had all gone home by the time he reached his office. The only interruptions were periodic visits from the night security man. Harper read over every cable and memorandum he had written from Tokyo, looking for the slightest mention of the code word *convergence*. He studied the surveillance logs. They were, as always, elliptical, edited by the slowness of the voice-activated switch, the hissing static on the line, the speaker's swallowed words. But what was inaudible did not matter. Harper was not looking for what was spoken. He was looking for what had been heard. Somewhere there may have been a breach.

The whole point of having the word was to exclude Langley and the station chief. He had devised it as a barrier against meddling, to protect his sole control over the operation so that what had happened in Vietnam could this time be prevented. Of course, Birch could simply have lied about the telephone call. But Harper gave him more credit than that. You don't write cover stories that conflict with the known facts. You fabricate around the truth, not against it. One way or another, he assumed, the word had been spoken.

By the time Harper finished his search it was dawn. The all-night announcer on the radio said it was a beautiful day. Harper slumped in his chair. He had found nothing. He had never spoken the code word to anyone but Birch, never written it down. It did not appear anywhere in the transcripts. *Convergence*. He was shaken by the way everything now seemed to point to him. He had recruited Birch. He had discovered the meeting at the Hay-Adams. He had conceived the code word. *Convergence*. But he had forgotten that convergence can go either way, rising or falling. And now he knew he was going to have to bring Birch down.

Miss Lutkin arrived, freshly powdered and creased, announcing that he looked as if he had been on a bender. He thanked her for her deep concern and dismissed her. With the office closed, radio off, and orders given to hold all calls, he paced off the ugly alternatives. What if Kerzhentseff knew the word? Birch had nothing of substance to offer him, and Kerzhentseff must have known it. Yet he had risked contact with Birch anyway. Had Zapadnya set up a trap to ensnare Harper himself this time, to get his revenge? Either in league with Birch or on his own, Kerzhentseff could have arranged it all. The telephone conversation. The clumsy contact at the Hay-Adams. The encounter at New Market, person to person against all the rules of tradecraft. Kerzhentseff would have known that the story about the code word would never fully absolve Birch. But he would also have known that the doubt it would cast on Harper would never wholly be cleared. The perfect last stroke of the satire, slicing through Harper's unprotected neck.

But still, how could Kerzhentseff have known the word? It came down to Birch again. If the Russian had managed to find out about it some other way, it would have to have been in Tokyo, where the word was used. But if he had known it in Tokyo, why had he let the operation go to a conclusion? Why had he shut Black Body down? It did not hold together. It was no more plausible than the other frightening hypothesis: that he was being set up from the inside. The evidence just was not there.

After the event, Harper was perfectly candid about having considered every alternative. It was his training, and he would have been suspect if he had tried to pretend otherwise. This was the most delicate moment, the moment of maximum pressure when everything closes in, when an agent can be driven to any extreme of conjecture and escape. He was not immune to the subversive thought, he admitted much later. He ruled nothing out, he insisted. Nothing.

The intercom buzzed. He let it squawk.

"I was ringing you," said Miss Lutkin, stern in the doorway. "He's on his way down here right now."

"Who?"

"The Deputy. You'd better get yourself together."

"Thank you, Miss Lutkin."

She did not even bother to sniff before slamming the door. And he did not bother to hike up his tie.

"Don't get up," said the Deputy as he stepped briskly in without knocking. "Odd business."

"How much do you know?"

"I'm not sure that is the pertinent question," said the Deputy.

"I was here all night going over the files," said Harper, as if diligence were the issue. A guilty man would work hard too, maybe harder. "I'm sorry to say that there is no way anyone else knew the signal. It was between Birch and me alone."

"Are you quite sure you told no one else?"

"I'm not sure of anything."

"You did not trust the distribution list."

"A precaution. No one else needed to know."

"We might have an interesting debate on that subject," said the Deputy, "some other time."

"I tried to be cautious."

"And you wanted to retain the ability to overrule any instruction that did not suit you," said the Deputy.

It was no great leap of induction. On the first level the thing spoke for itself. It was one of those transparent devices field agents use against their superiors. Very common. Beyond that, the interpretations could diverge. Harper did not leave the worst unspoken.

"I would say that by the process of elimination I am now the focus of the inquiry. If I were dispassionate, that would certainly be the approach I would recommend. And it follows that I should be removed from the case and isolated, perhaps put on administrative leave—paid, of course—pending further developments. Let the suspect sit and stew for a while. I will, of course, cooperate fully. And I will accept whatever judgment is made."

"Don't give me a speech, Harper," said the Deputy. "It gets on my nerves. Makes me think there's a recorder grinding away somewhere. There is no posterity here, so there's no point preparing your case for it.

"We have discussed the matter fully, and the consensus is that you should not be relieved of your responsibility. Mr. Birch has put you on the spot, but we don't let rogue assets force decisions on us."

"I would like to think it was only Birch gone bad," said Harper.

"Odd," said the Deputy. "I might have thought you would have lost some of your famous confidence in him."

"Maybe that's too easy," said Harper.

"One never knows when the most obvious is also the most subtle. I suppose you will want to interrogate him."

"I will."

"The way is clear now."

"The Sisters botched it. He's fair game."

"There are certain things we can always count on. It is quite a solace really. Especially when the flap potential is so large."

"Are the Sisters investigating me?"

"It would be somewhat indiscreet to ask, don't you agree?"

"I have to assume that they are."

"Over every shoulder there is someone peering," said the Deputy. "We march in one great suspicious circle."

"Especially when something goes wrong."

"Your nemesis, Mr. Kerzhentseff, could not have done better if he had tried," said the Deputy.

"It's good of you to mention that."

"Don't overestimate the depth of my preconceptions," said the Deputy, smiling. "I rule nothing out."

"An open mind," said Harper, with an unfortunate edge.

"Which is not," the Deputy added, "the same thing as tolerance."

The plan came upon Harper suddenly. And as he went through the preparations, it came together so easily—down to the right personnel just happening to be stationed at Langley—that he could not help entertaining the question of whether he wasn't following a trail cut for him by others.

Birch was most agreeable when Harper telephoned him. He was feeling a little frail from the night before, he said, but he was willing, even enthusiastic, about the chance to be interviewed, to clear his name. Harper, of course, did not lay out all the details. If his scheme had any value, it was in the element of surprise. They agreed to meet the next day. Harper said he would have him picked up at his apartment. Birch suggested a more evasive rendezvous, but Harper assured him that it was not necessary anymore. When he hung up, Harper had Miss Lutkin transcribe the tape.

With all the preparations underway, Harper could not hold the fatigue back any longer. He found himself drifting off into the bizarre connections of sleep—Birch and Kerzhentseff merged, Kerzhentseff and himself. And what troubled him was that the dreams did not seem implausible. He left the last of the details to Miss Lutkin and drove home.

It was still before noon when he fell into bed in a deep and mercifully memoryless sleep. And when he awakened with a nervous start that was itself a kind of memory, he was disoriented by the light. The clock read 7:15, but it had to be wrong. He could not imagine that he had slept through until morning. It was only when he turned on the radio next to the bed and the soothing, distant voice gave the time that he realized that the clock was right and it was evening.

Now that he knew he did not have to rise, he was painfully awake. He rolled his eyes up under the closed lids to the aching spot. He tried to breathe himself relaxed, but every muscle was taut. He showered and shaved, feeling lagged as a traveler on a new continent, and went downstairs.

A bacon sandwich was a decent equivocation between breakfast and dinner, and though he poured himself some wine, he found he had no taste for it. The Beethoven quartets were still on the turntable from his interrupted evening with Fran. He lifted them on the spindle and switched the machine on. The music was rapturous but ordered, both at once, but without duplicity. He knew that he needed her.

He dialed Fran's number but was not able to outlast her patience in letting it ring. He knew that she was there, as if in the hollow space between the buzzes he could hear her soft breathing. She was still punishing him for the love of his wife. No, it was not really the love. She had always assumed and accepted that. He had wronged her only by his failure to keep that jealous affection hidden. He had failed her because secrets are to be kept, not shared. They do injury in the light. He turned off the record and left the house.

The pavement was slick with rain, and the traffic down Wisconsin Avenue was even slower than usual. At Q Street a limousine with diplomatic plates ran the light in front of him. Two men in the back wore turbans and tuxedos. They sipped wine and did not even seem to hear the shriek of Harper's brakes. The young people and their imitators were out in Georgetown despite the rain. They paraded hand in hand—women and men in couples of all combinations. At the corner of M, the Hare Krishnas danced, saffron dervishes, for a small audience of tourists. Harper turned right and inched with the traffic toward the bridge.

Up ahead, across the river, the tall office buildings of Rosslyn rose in an ugly disharmony that could only have been accomplished by plan. Harper drove quickly through the deep canyon between the buildings, then swung his car down the ramp to the parkway. An airplane was making its final approach to National Airport above him. Its lamps cut long beams in the mist and lighted up the crowd of people that gathered on a spit of land just short of the runway to risk the near misses of the jets' descent. Harper saw the plane touch down in a cloud of sparkling water and then relax all at once, wings sagging, like something very old.

The traffic was much thinner in Alexandria. For some reason the historic section, even redone, had been spared the popularity of Georgetown, at least for the moment. He turned away from the landmark neighborhood, drove down a hammering brick road, and found a place to park at the curb a few steps from her building.

When he reached her door he heard music, drums, and voices. It was something African, sacramental. He knocked loudly, out of cadence. The music softened. He heard movement inside, the sound of glass against glass. Then the door opened.

"There is a reason I haven't answered the phone," she said, holding the door close to her body, blocking his view inside.

"I didn't realize there would be someone else here," he said. "I wouldn't have come."

"That would make it easier for you, wouldn't it," she said, swinging the door open on the dim and vacant apartment. "I'm quite alone."

Her bare feet gripped the metal baseplate of the doorway, and the cutoffs made her tense legs seem long against the cold steel of the frame. Her hair was pinned up and careless. He could not believe that she had not expected him.

"I came to make amends," he said, hoping by his stiffness to make her laugh. The primitive music beat away, and she was not amused. "It was a terribly timed coincidence," he went on. "Worse was how I handled it. Could we talk?"

She stepped aside to let him in. It was acquiescence, not an invitation. He found the couch and left her a space. She went down Indian-style on the floor, putting the coffee table between them.

"I would have come looking for you sooner," he said. "But I couldn't break away."

"Held up by all those terrible secrets," she said. But it was not one of her taunts this time. It hurt him.

"I would share them," he said, "if I could."

"What exactly was it that you wanted to say, Richard?"

"I could use a drink," he said.

She got up and returned with two glasses. Hers was wine, his whiskey and water. She sipped her drink and brushed a loose strand from her face. She did not, he supposed, intend it to be alluring.

"Would it make a difference if I explained?" he asked.

"No," she said. "I'm young. I learn."

"And I am older," he said. "I've made more promises. They hedge me with obligations I couldn't deny even if I wanted to."

"It's not your having a wife," she said, "if that's what you mean. I could tolerate that. Have."

"But not so explicitly," he said. "I understand the difference. The other night, . . . I should have let the telephone ring. You could teach me the discipline."

"I don't want to be with you anymore, Richard," she said.

And he knew that it had to come to that. He had anticipated it, a crisis that had to be met and passed. Yes, she would have to remind him of the consequences. Then, yes, he would take seriously what he could lose. And, yes, he would show her that he feared it.

"You have every right to feel that way," he said.

"It isn't a question of rights," she said. "It's what you are, what I've finally seen."

"A hypocrite?" he said. "I did not lie the other night. Not to either of you."

"God, if you could have lied, then maybe it would have been all right. But you don't lie. You are what you are, and you don't pretend otherwise. It's all right there," she said, her arm outstretched, pointing at his weak places, which were everywhere.

"I haven't performed very well lately," he said.

"It isn't anything that easy, Richard," she said. "Why is it that men always think that is the problem?"

"Maybe they're afraid it is the one problem they can't solve," he said.

"I can always have my pleasure," she said. "It's the least discriminating part of me, Richard. All it knows is hard and soft."

"No small difference," he said, but his mockery didn't help.

"Small enough not to matter," she said. "I thought you had a passion. I thought it was something you felt you had to hold out of sight. But I thought that it drove you, a secret memory, a secret purpose."

"I have too many."

"What I heard the other night was your secret side," she said.

"It was only my wife. I never hid her from you. I mean, there are things I can't tell her, either."

"About me, for example."

"About work. The things I do, am."

"I heard your secret, and it was nothing. You don't believe in it so deeply that you would risk everything for it. You don't risk anything. You just want your comfort, to keep things orderly, lined up like columns of figures. You can calculate that way, Richard, but you'll never get beyond sums."

"I don't know what you mean."

"What you hid from me was nothing more than a man trying to save all his pleasures. The secret was just how far you would go."

"What's wrong with being like everyone else?"

"There is nothing wrong with being just another GS whatever, holding on, just another saver. There's nothing wrong or right about it. It's empty, that's all. For me it isn't enough."

"You wanted a hero," he said.

"I wanted to believe in something that no one dared to name."

He waited silently for a moment, watching her. She was not about to give way.

"I am in trouble, Fran. Serious trouble," he said. It was a way of pleading, and he knew it was pathetic. But he was willing to do nearly anything now to touch her. He buried his head in his hands so that she could imagine at least the tears.

"It will turn out fine, whatever it is," she said. "You will manage."

"I want you," he said. She stood up, pulled down the tattered legs of her cutoffs.

"Look," she said, "if we are talking about going to bed one more time to say good-bye, that's OK with me. But I had more at stake before. And now I don't anymore. So either way, it's all the same to me."

"I guess I don't know how you can be so cold-blooded about it," he said.

"Put your mind to it," she said. "I think you can."

30

Purists would say we had no business taking the chance we did with Harper. He was under suspicion, and the strain was increasing. His whimpering conduct with this Larsen woman was appalling. It is one thing for a man to seek pleasure, even to pay, but it is quite another thing to beg for it. This and the fact that he himself had begun to figure into Birch's allegations led some in the Agency to insist upon appearances.

It could be argued that the Deputy and I made a bad calibration in letting him put more pressure on Birch. No one regrets more than I do the misfortune that developed. But if our deliberations appear bloodless, this is only because one can never adequately predict the sanguine element. It was not, I assure you, that we failed to appreciate the possibility of error. No one knows better than we do at Langley that one cannot always split the human diamond precisely on the flaw. But Harper had a plan. And it was worth the risk to allow him to put it into effect.

We did take certain precautions so that the record, which up until this point had been riddled with fissures and ambiguities, would henceforth be ironclad. We recorded every word, of course, and made a complete transcript. This proved to be quite helpful later when I took Harper back through the final interrogation of Birch second by second in an effort to make sense of the events it set into motion. But I did not have to rely entirely upon Harper's reactions to the naked words on paper and tape, his recollection of Birch's gestures and expressions. With the concurrence of the Deputy, I observed the questioning. I was there.

The room Harper set up at Fort Belvoir was not an exact replica. There had been no photographs taken at Camp Zama, for security reasons, so they had to set up the electronic array and stack the empty boxes by memory. The walls may have been a slightly different shade of beige. The building was not quite as squat and not nearly as segregated as the headquarters of the dummy unit behind its high wire fence. But the idea was for it to be suggestive, like a stage set, and the recognition registered immediately in Birch's expression as they led him inside.

"I think you know nearly everyone," said Harper. Even the silent men he had detailed to pick Birch up at his apartment had been on the Tokyo surveillance team. There was always a chance that at one time or another he might have noticed them. The polygraph operator, as always, was busy at his dials and did not even deign to look up as his old subject came haltingly forward.

"What is this supposed to be?" Birch demanded.

"You are thoroughly familiar with being fluttered," said Harper. "We wanted you to feel at home."

No matter what in the end Birch proved to be, he was essentially a simple man, astonished at the pains guile took against him. As sure as Birch himself was in duplicity, it was merely an instinct. And in the face of a less natural contrivance, he seemed startled and offended. Harper motioned him to his familiar chair next to the machine.

"Can I talk to you first?" Birch whispered.

"I'd frankly rather have you on the polygraph," said Harper. "It's cleaner that way. The record is more complete."

"Please," said Birch.

"All right. Over here. The rest of you get ready. We won't

be long." Harper maneuvered him to the corner where we had the microphones installed against just such a contingency. As a fallback, Harper himself was wired for sound. But the reception from body mikes was always fickle. You might find yourself recording the heavy luffing of your own pulse or the rude scratch of a button against the lining of your suit. We did not want to risk losing any of the precious, potentially revealing words. And he could not afford the accusation that behind the noise he had somehow negotiated a separate peace.

"I don't get this," said Birch, "all this past."

"Doesn't it amuse you? I thought it might. Old times."

"What do you want from me?"

Harper laid his hand gently on Birch's shoulder and watched him flinch. He took the hand away. The hidden, overhearing ears helped him keep his distance. The trick had to be played out to the end.

"It isn't a matter of what I want," said Harper. "There are just some questions."

Birch had reason to be upset. With every phrase, Harper accentuated the differences between them, differences of rank and class, differences of authority.

"Say the word," Birch pleaded. "Say the word and tell me what I'm supposed to do."

"I guess the word is part of the problem, Birch."

"When you said it, I knew everything was all right," said Birch. "It *is* all right, isn't it?"

"You are on your own, my friend."

"Convergence," said Birch. "It was the signal that the orders were valid. Remember? Simon says. And it was the signal that I was in trouble. I was only to use it in an emergency. Well, I'm using it now, Mr. Harper. Convergence. I don't know what you want of me."

"Just answer the questions, Birch," said Harper in as bored a voice as he could manage. "Tell the truth."

As Birch looked at him, all the mazy lies to which he had been exposed showed in those pathetic, vacant eyes. He was for the moment nothing but a cipher waiting to be assigned its meaning, a mirror ready to repeat whatever image presented itself. Birch was a secret whose secret might have been nothing more than that it was a void.

"Just tell the truth," Harper repeated. "Come on, now. We've wasted a lot of time. We have to wire you up."

"Will you believe me then?" Birch said. Harper did not answer. "You always said I could beat the machines."

"Yes, I did," said Harper.

"And I thought you meant it."

"Did you?"

"Kerzhentseff never found me out. I fooled his machines."

"That is what we are here to discover, Birch," said Harper, turning his back on him, "just who has been the fool."

Of course Birch had located the problem. Harper needed to know whether what was tormenting Birch was a sense of betrayal or discovery. Against a man as talented and trained as Birch, the most delicate instrument of silicon and steel would not be capable of the distinction. Harper had to rely on the more sensitive mechanisms that operate between man and man. He had to bewilder Birch, snag him up in his own vagrancy and doubt until his falsehood, if that was what it was, collapsed in on itself.

Birch gazed out the window into the middle distance where the landscape was familiar. Harper noticed this and had the glass covered with a colorless cloth. If it irritated Birch, he concealed it. The pens etched long and gentle curves.

The preliminaries—name, age, occupation—went smoothly enough. Too smoothly for Harper's design. He had hoped Birch might be shaken, might try to resist him as he had that last time in Tokyo. But Birch was steady, and Harper knew it would take a mean persistence to break him.

Q. Are you acquainted with a code word I used in Tokyo?

A. Yes.

Q. Was this a special sign between you and me alone?

"I'm sorry. I can't answer the question that way."

Q. Did I tell you this was to be a sign known only to you and to me?

A. Yes.

Q. Did you use the code word in Tokyo?

316 CONVERGENCE

A. Yes.

Q. Did you identify it as a way of knowing and being known?

A. Yes.

Q. When did you first tell Kerzhentseff of the signal?

"I'm supposed to answer yes or no," said Birch. "Those were the rules. You asked me a question I can't answer that way."

"You can't answer it?" said Harper, glancing knowingly at the technician, who made a red mark on the paper tape.

"Not yes or no," said Birch. "The question was when I told Kerzhentseff. I can't answer when."

"Don't you want to answer it?"

Birch turned in his chair, and the delicate pens went frantic.

"I'll answer anything," said Birch. "But I want to follow the rules."

"I suppose the rules are a great comfort to you," said Harper. "They give you a chance to relax, the way I taught you. To buy time."

Birch sat back, bringing the needles under his complete control. His breath was calm and regular again, his forehead free of any line that might be read.

"I never told anyone about the code word," said Birch, and then he glanced over at the pattern his body transmitted.

"But, of course, you did," said Harper.

"No," said Birch.

"Two days ago," said Harper. "You can remember that far back, can't you, Birch? You told agents of the FBI the word in my presence. I don't understand what you think you will accomplish by this lie."

"Well, you knew it," said Birch, starting forward. "That wasn't what you were asking about because you already knew it."

"Be still," the operator commanded.

Harper sat impassively, his fingertips touching in a cathedral of rectitude.

"Yes, I knew it," he said, "just as I know your name and age. But you should not tailor your candor to what you think I know. That's my advice to you now, Birch. You cannot be sure of what I may have learned."

"I didn't realize you were trying to trip me up."

"You are very sensitive about this point, aren't you," said Harper, rolling the scrolled paper in such a way that Birch could not see what he was finding there.

"I never told Kerzhentseff about the code signal," said Birch. "Did I pass the test?"

Q. Have you told anyone else?

A. Yes.

Q. Have you told the FBI?

A. Yes . . . But only because you said I should.

"That's all right, Birch," said Harper. "You needn't be defensive about it. I wanted you to cooperate. There are no secrets between the agencies. We have a common goal."

"All of us do," said Birch.

"The Agency and the Bureau," Harper corrected him. He waited before beginning again. There was a danger in working close to the edge so soon.

Q. Have you told anyone other than the FBI?

A. No.

Q. Have you mentioned it to your wife, perhaps?

A. No.

Q. Maybe someday when you were a little drunk. Isn't it possible that you might have let it slip? It would be natural, with a loving woman like her.

"I don't get drunk often," said Birch. "It only happened that one time. I'm sorry about it. But that was the only time, and that was after."

"Sometimes it is difficult to remember what one says or does while drinking," said Harper.

"I don't remember not remembering," said Birch. "How can I be sure of something you don't think I can recall?"

"You are hedging, Birch," said Harper. It was a trick as old as orthodoxy: to snare the suspect in small heresies, paradoxes of dogma that even the prophet could not reconcile. You wrenched out admission after trifling admission until they began to pile up and make the larger denial seem suspect. Yes,

Harper was part of a grand tradition. He worked at the end of a history that reached all the way back to the Inquisition and beyond. And it gave him no consolation that all the unfairnesses of the past, whatever their tribute of blood, had always served the historical imperatives—the rise and fall of civilizations and faiths, the corruption of men and societies destined to die.

But I knew from looking at Birch that, by themselves, Harper's tricks would never have caught Birch up. He was too good, and they were far too crude. Kerzhentseff had been a more worthy opponent for the young man. Of course, in the end Harper took responsibility for the consequences of the interview. I have no doubt that the pangs this gave him were no less real than the pride. But he need not have indulged in either guilt or satisfaction. What happened to Birch, Birch did to himself.

"I don't want to hedge," Birch said. "I want to be complete. I did get drunk that night. And you know why."

"Do I?" said Harper.

"How am I supposed to answer that? Yes, you know? No, you don't? I have no idea."

"You seem to think you do, Birch," said Harper. "You seem to be trying to trim to the wind. But be careful where it leads."

"Look," said Birch. "I heard the code word on the telephone and I trusted that one way or another it came from you. I trusted that because I had never told anyone else. And now you say that it wasn't you. Maybe you are the one who should be hooked to the box."

"I'm sorry if your faith has been shaken," said Harper, coldly. "Let's get on to that telephone call. You seem eager, so let's discuss it. I should have thought that you would want to ease into the subject, since it is the one where you are most vulnerable."

"I'll talk about anything you want, Mr. Harper," said Birch. "And I want you to know that I am not afraid."

"That has always worried you, hasn't it?" said Harper.

Q. On the evening of the twelfth, did you receive a telephone call?

A. Yes.

Q. Under what circumstances?

"Do you want me to answer that?"

"Of course."

"I wasn't sure. I try to do this the way I am supposed to."

"And I am trying to make it as simple as possible for you. Go ahead."

"I was asleep," said Birch. "The ringing woke me up. Donna went to answer it then came back into the bedroom to say it was for me. When I picked up the receiver and asked who it was, the man on the other end said the word. I asked him to repeat it, but he didn't. He seemed to be in a hurry. He went right ahead to say that I should make a drop behind the Hay-Adams on a particular night. I was to write a note identifying my current assignment. The note was for Kerzhentseff. The way the man on the phone put it, I was to be in contact with the Tokyo mark again. That was how he said it. The Tokyo mark. That was all. He hung up."

Q. Were you tired when you took the call?

A. I was asleep. I mean, yes, I was tired. Yes.

Q. Are you sure you weren't dreaming?

A. Yes.

Q. But you did not feel completely alert.

"I was awake. The next morning I asked Donna. She remembered the call. And I found the piece of paper where I had written the place and the time he had set for the drop."

"You felt the need to ask your wife about it?"

"I wanted to give her a cover story."

"Did the man speak the word to her?"

"Of course not. She doesn't eavesdrop. She went back to bed, and by the time I was finished she was almost asleep again."

"On the piece of paper, did you write the code word?"

"I didn't think it was something I should put down."

"But you did not trust your memory with the other details."

"It was late. I wanted to be sure in the morning. You know how you are sometimes at night. It was better to have it written down."

Harper stood up and paced. Birch followed him with his eyes. Harper obviously wanted to convey the sense of a hidden detail, a fact that Birch was contradicting without knowing it. No matter how attentive you are to the small points in a lie, you can never be sure that it will not be snagged on something peripheral. It was an inquisitor's advantage; against a man like Birch, it was a small one.

Q. You asked the man on the telephone to repeat the word?

A. Yes.

Q. You did this because you were unsure about what you heard?

A. No. I heard it.

The machine registered a perfectly calm response, but Harper disregarded it. He moved in quickly, as if he knew this was the weak point. It had to be. It was the point that impaled Harper himself.

"You were sure you heard it, but you did not write it down. You were sure you heard it, but you asked that it be repeated. It is a funny kind of certainty, isn't it?"

"It was a relief, Mr. Harper. I'd wanted for so long to hear from you again, to get another assignment."

"You wanted it repeated because you can never be told too often that you are needed and loved."

"Yes. That's what it was like. I was very happy."

"Don't you think you are pushing it too much? I thought I had trained you better."

"They were your words, being told that you are loved. It was something like what I felt."

"And you thought that if they were my words, I might find them convincing."

"I thought you might really be trying to understand. You never acted like this before. Even when you were playing the Russian's part during training. I don't know what you are trying to do."

"We have to hold some techniques in reserve, you know," Harper said. He leaned over and flicked off the polygraph, startling the technician, who was busy marking the places where the unorthodox questions and answers began. The man moved to start the machine up again, but Harper stopped him.

Birch was shattered now, and it took all the dedication and fear that Harper could muster to press on.

I had seen this sort of thing happen before, a man finally breaking under all the lies that weigh down upon him. Often no single factor is decisive. It is, instead, the mounting inability to separate one set of lies from another, let alone the lies from the truth. A man like Birch, prodigious as he was at the game and capable as he had demonstrated himself to be at improvisation, had always in the past worked from a thoroughly crafted text. He had rung changes upon it, of course, masterfully so. But now he was working without a script. There had been no opportunity for anyone to prepare him. He was utterly unmoored, and all the falsehoods he had lived buffeted him without system or surcease. Like the machines and circuits Birch had learned to riddle, his very mind itself had finally succumbed to his ability to deceive.

"I beat the machine, didn't I?" Birch asked.

"Yes, you did, Birch."

"I'll never make you believe me now."

"You have been lying," said Harper.

"I don't know," said Birch. His body, encumbered by the wires and straps, collapsed into the chair. He was too tired to defend himself anymore. His eyes were glazed with tears, mirrors that obscured him, held him apart, alone. "I just don't know."

"You never heard the word," said Harper, softly.

"If it's better that way, then I didn't. If that is what you want me to say."

"I want you to be candid."

"Then all I can tell you is that I don't know. You taught me how to remember only what I was supposed to remember. I wanted to hear the word. And wanting was enough. It is the only memory I have. I wanted to be in the action again, and I would have done anything to get there."

"Even work for the other side?"

"Not that," said Birch.

"But not to care who you were working for."

"To care," said Birch. "To care so much that I only know how much I wanted the man on the phone to be you. God, I wish the machines could tell me the truth too. I wish there were a way of being sure."

"So you admit," said Harper, for the record, for himself, "that it is possible that you did not receive the signal. You admit that it is possible that you went off on this thing without having any idea where it would lead you or even who was doing the leading."

"It's possible," said Birch. "Don't you understand? That's the sheer hell of it."

"Possible," said Harper. "But is it true?"

"I am not a traitor."

"Perhaps we have had enough for today."

"I've got to convince you of that. I am not a traitor. Just that one thing. It is in my heart. It is sure."

"I'm afraid your heart will have to wait for another session," said Harper.

"I know what I intended," said Birch as he stood, undoing the straps and peeling off the electrodes. He gazed at Harper as if he still expected to be given the sign of identity that could not be given. I watched him standing there as frail and powerless as the truth.

31

The red brick apartments squatting on their path-worn lawns of brown had never looked to Birch more like barracks. At his request, the escorts had dropped him off a distance from home. He told them he needed to pick up a few things on the way. In fact, he just needed to walk and sweat and tell himself that he was free to go where he pleased and be whoever it was he had become. But it was no use. Everywhere he looked it was the same. He had nowhere else to go.

It was no use thinking either. The memory was not going to become any clearer. He could not summon it up like a word on the tip of his tongue. He knew exactly the word he wanted. But he did not know whether it had been used. Tomorrow there would be more questions, and the next day the same. He would be seated, arms flat on the rests, palms down. They would wire his wirsts, strap him. The machine would sit there humming away, the operator dutifully marking the paper tape, and it would mean nothing. He had learned his lessons well in Tokyo because he had thought it was his duty. Now he found that the training sealed him off. It was worse than just being on his own. He was walled out of any possibility of trust. Birch, the perfect liar. Birch, the cipher. Birch, the fool. He did not even know himself.

Birch mounted the stairs of his building heavily, stopping at the open landings—balconies overlooking other balconies—to get his breath and to feel for a nonexistent breeze. As he entered the apartment he noticed that the air conditioner had provided only a few degrees of coolness. The air was as dank as a cave's. It smelled musty and mildewed, confining. It was a cell.

"Donna!" he called. She emerged from their room and managed a smile.

"How did it go, Jerry?"

She was timid of the question, as if she only asked it because this was expected of her. She stood away from him, and he did not move to meet her.

"Air feels rotten," he said.

"The weather has been just awful."

"You could have turned it on earlier, while I was gone," he said.

"It's been on all morning, Jerry. There's only so much it can do."

"What did we buy it for if all it does is make the place smell like a goddamned outhouse?"

"I'll turn it off if you'd rather open the windows." She started toward the straining unit.

"Leave it," he said. "Leave the damned thing on. I was just hoping it might be pleasant, that's all. I'm stuck here, and I thought at least it might be comfortable."

"I'm sorry, Jerry," she said. "It didn't go well, did it?"

"How would *you* know?"

She backed away from him.

"Can I get you something cold?" she said. "I think there's some soda left. I could mix up some iced tea."

"Not now," he said.

"It might make you feel more comfortable."

"Not now!"

His voice was much too loud. It startled her. He had not meant to make her afraid, yet it gratified him that she was. She

damn well ought to be afraid, even if he couldn't tell her exactly why. They were in trouble, and she might as well get
used to it. As he took a sullen spot on the couch, she slipped
quietly out of the room.

The newspaper was there rolled up next to him. He strug-
gled with the rubber bands that were twisted tightly around the
middle. They would not slip off, and he pulled at them in a
frenzy until they broke. The paper tore under his fingers, wet
and soft. He flung it away.

"Donna!" he called out. "Where did you go to?"

"I'm just doing some sewing in the bedroom," she said,
barely audible. "I thought maybe you'd want to be alone."

"Get in here, will you?"

She dutifully joined him, looking tired and old. A piece of
fabric hung limply from her hands, dangled with bits of thread.

"You shouldn't just turn away from me," he said.

You could only make contact by inflicting pain. It was a
kind of belief, something you could see. She sat down next to
him on the couch, and he looked critically at her old clothes
and careworn figure. She could at least try to make herself attractive to him. She could do that much.

"I didn't think I could do anything for you," she said. "I
wasn't being of much use. I had things to do."

"You can do your knitting in here."

"The light isn't good. And it isn't knitting. It's embroidery.
For your cousin Joe's new boy. Did I tell you we got an an-
nouncement?"

"No," he said. "No, you didn't."

"It came just the other day. I should have mentioned it. I
was going to. But you seemed so . . . so busy."

"It's my own family. I ought to know about my own family."

"I'm sorry, Jerry."

He hated her apologies. They made her so small and weak,
him smaller still.

"I want to know these things," he said.

"I want to know things too," she said. "You only tell me a
little. You keep the rest all pent up. I need to share whatever it
is that is happening to you."

"I'll tell you what is happening," he said. "I'm all fucked up."

He did not use that word around her, did not tolerate others using it either. And now it fell upon her like a blow.

"It must be very bad," she said softly.

"They don't believe me," he said.

Maybe if he could say it out loud, even part of it, he could master it. Maybe she could help, or at least she could join him in his glass cage.

"They think I was working for the Russians," he said.

"Oh no, Jerry. You weren't, were you?"

"It doesn't matter," he said. "It's very complicated."

"Of course it matters," she said, touching his hand with fingers that burned him. Her trust was as automatic as Harper's doubt. It was meaningless.

"You don't know a damned thing about it," he said.

She withdrew her hand and studied her lap. It was a relief for him to be free of her touch. She could not come inside with him because there was no room for her. They could only flail about hurting one another.

"I don't know," she said, "because you won't let me."

"To keep you safe," he said. "That's why. It's not for me. It's for you."

"If you are in danger, so am I. I'm with you no matter what you do."

"Were you with me when I went up against the Russian in Tokyo?" he said, leaping from his seat. "Were you with me this morning when they accused me of being a traitor?"

"Yes," she said and then she took up the needle, trying to work it tightly through the cloth. Her hands shook, and Birch saw the steel point as it found her flesh. She closed her eyes but did not cry out. A tiny droplet of blood beaded on the skin. It inflamed him. He swept the fabric and thread from her lap to the floor.

"Put that stuff away," he said. "You're going to ruin it. Listen to me when I am trying to talk."

"I'm trying to listen, Jerry. But sometimes you make it so hard." Her tears incited him. They closed him out. They silenced him. He had protected her as long as he could, and now she was protecting herself from him.

"Stop it," he demanded.

"I can't," she said. "I can't help it."

"Stop."

"You're somebody else, Jerry." The tears welled from her ugly, reddened eyes. "I don't even know who you are anymore."

He pulled her roughly to her feet. She came up obediently, and for an instant she leaned forward as if to embrace him. But he was in a fury. He was on the inside and she was on the outside. She was too scared, too selfish to join him against the world. She was deserting him, just as everyone else had. She closed him out and said it was because he wasn't who he was. Well, who the hell was he supposed to be? They didn't tell you that. None of them did. They were all the same. They used you up and then ran away. And you are supposed to be the one to blame because nobody tells you who they want you to be.

When he slapped her, open-handed and hard across the mouth, she went down and curled against the next blow.

"Get up," he said. She cowered below him. "Get up!"

She turned her head up to him, the blood smeared on her mouth like the lipstick of a whore. Then she began to crawl to the door.

"Don't go," he said.

But she left him standing there in rage and remorse, not knowing which was more real.

He did not follow her. He did not trust himself anymore to know even his own heart. He had injured his wife. It had come to that. He knew what he had to do.

Birch went to the desk in the spare room, drew out a single sheet of paper, and as he had when he penned the first message to Kerzhentseff, he wrote the words in an unconscious rush. There was no time now for reflection or doubt. If you have a mission, you have to put all that aside. It has to be instinctive, unwilled. To try was to fail.

When he finished, he folded up the paper and, without bothering to seal it in an envelope, put it into his pocket. He

placed the telephone call quickly, his fingers hurrying the dial.
The familiar voice answered, and Birch said no more than was
absolutely necessary.

When he got outside, the heat did not oppress him anymore.
He barely noticed the spots of blood on the hallway tile that
tracked Donna to the safety of their neighbor's door. He made
it to his car at a half-run and drove without looking behind.
Let them follow him; he would make this rendezvous no matter who knew it.

32

She had already stopped her crying when she heard the door slam in the hallway. Donna held the washcloth-wrapped ice cubes to soothe the swelling lip and listened to his footsteps on the stairs.

"It always comes down to brute strength," said her girl-friend. "When they're most in the wrong, they go with strong."

"Lend me your car," said Donna, conscious of the way her numb lips misshaped the words.

"I'll take you to the emergency room if you want."

"He's going somewhere," said Donna.

"I don't think the cut is that deep, but if you think you need stitches . . . Here, let me look. It seems to be all on the inside. There will be a bruise. I suppose he'll be proud of that. But the scar won't show. He'll be glad of that too."

"I've got to follow him," said Donna.

"The hell you do."

"I need your car."

"The important thing is that you can't want to be hurt. Some women do. You can't let that happen to you."

"He won't hurt me again."

"I don't know," said her girlfriend, fishing the keys out of her purse and handing them over. "If you want to do this, it's up to you. You have to kind of jiggle the key in the starter to get it to turn over. Here, let me drive you."

"I'll drive myself," said Donna.

"Be careful."

Jerry was pulling away far down the block by the time she fooled the ignition into action. She lost sight of him around the corner but picked him up again in the lazy midafternoon traffic. Donna kept her distance, trailing him down the long, ugly corridors of Arlington. She couldn't let him see her. If he was going to a bar, she would let him get inside and calm down a little first. Then she would go to him. If he wanted to drink, she would drink with him. If he would come home, she would welcome him. He must have hated himself very much to have done what he did to her. Jerry was as lost to himself as he was to her, and she had to find him and bring him back.

But where to look? She did not even know for sure what evil was upon him. The CIA, the Russians, secret missions in Tokyo. She had been so blind. She had never been able to share his glory, and now he was trying to keep her safe from his despair. Let me in now, Jerry, she prayed from three cars' distance. Make a place for me in your pain.

He turned onto Route 1, and she followed him mile after squalid mile. Fast-food joints, discount stores, car dealers, sleazy motels, taverns. Up ahead his car pulled off the road. She was caught behind a Chevy turning left and could not see what the place was. There were bars along this road, but she could not imagine that was where he had stopped. They were so cheap.

The traffic cleared, and the Chevy made its turn. She sped forward and eased into the right lane, but when she reached the driveway where he had gone, she could not bring herself to stop. It wasn't a bar after all. It was a motel, and their car was parked there in the gravel. She caught a glimpse of Jerry's red shirt as he disappeared into the office.

It was a decrepit place, and the sign in front advertised rooms by the week, day, and hour. It reminded her of that awful place outside Cleanthe, The Drop Inn, where all the filthy boys and their fathers went on Saturday nights. She drove past and stopped at an intersection. Why this? Get drunk. Even strike her in a mindless rage. But she had always counted on his loyalty. She had been so sure of that. Suddenly she saw that it was this that had drawn him into his trouble. This was what corrupted him. It made her feel violated even to think of it, dirty inside.

She turned at the corner and wound her way through the back roads for a second look. When she reached the motel, he was at the door of one of the rooms, working at the knob. She slowed down. Someone honked a horn behind her. Maybe he would turn and see her. Maybe it would bring him to his senses. He did not turn. The door swung open into the darkness. And he was gone.

Donna drove on, turning on a highway that led home. She drove blindly on through her tears, wanting to turn back, wanting to turn back to the very beginning, before any of it went wrong. Go to him, she heard her friend's voice imploring. Go to him no matter how bad it seems. But what could she do? He was leaving her, leaving for some other. Then the same voice changed its tone. Leave him, it said. He is lost. Protect yourself. Nobody else will.

In the end, after all the tears had stopped, it was neither voice that finally moved her. It was something more intimate and sure. It told her that he was in terrible danger. She felt this knowledge run through her, a chill that began in her wrists and reached upward to her spine. She pulled into a parking lot and turned the car around. Wildly she drove toward her husband, frantic at how far away he seemed. She came to a skidding halt in the motel's gravel lot and flung herself from the car.

When she reached the motel room door, she put her ear to it. If she heard the woman's voice, then she would know. Was it gentle? Was he cruel? Was the thing impersonal or something unspeakably deeper and more malign? All she heard was the hum of traffic and the hot wind, the false sound of the sea in a shell. She knocked, softly at first, then louder and louder when there was no response.

"Jerry!" she shouted. She would take him back, no matter

what, back to Cleanthe where he could simply be again what he was, what she loved. "Jerry, please come out."

"Hey," said a voice behind her. "Hey, what's the deal?"

He was a young man, fat and bearded. His T-shirt was spotted with stains, and the words said VIRGINIA IS FOR LOVERS.

"The man in there," she said, "he's my husband. Jerry Birch." She clawed through her purse until she found the little driver's license with her picture and name. She showed it to him.

"I don't know nothing about it, lady," he said. "He's just another dude to me. We takes what comes."

She hated his sneering smile, his dirty mouth. But she did not let it faze her. She had one purpose only.

"You have a key," she said. "Let me in."

"It's his room. Mr. Anderson. He's got it the day. Now you want to leave?"

He took her by the elbow, but she twisted away and faced him.

"I have to be with him."

"No fuckin' way, lady," said the man. "We got our rules."

"In the name of God," she said. "Please."

That name meant no more than any of the others, and he moved toward her and took her arm in his greasy fist. But then suddenly he let go. A car careened into the lot, spitting gravel dust.

"What the hell, lady. You call the fuckin' cops?"

The car door swung open and she recognized the driver immediately. She had despised him when he had come over the night Jerry was drunk. But now she ran to him.

"He's in there," she said. "Number Eight. I saw him go in."

"You'll have to open that door for us, friend," said Harper.

"Like I told the lady, no way."

Harper reached into his breast pocket and pulled out a slim leather case. He hung the identification close to the young man's nose.

"I got to get the master key," said the young man, moving away from them.

"He telephoned me," Harper said. "He gave this place and time. What did he tell you?"

"Nothing," she said. "Absolutely nothing." The chill had given way to numbness. She wasn't afraid anymore. It was lost.

"Maybe you'd better let me talk to him first," Harper said when the young man ambled back with the key.

Harper opened the door, and the room was empty. The bed was unrumpled under its cheap cotton spread. There was no woman. A little television set sat bolted on a rickety shelf. The room smelled musty and sweet. The carpet was shag, worn bare in a path from the doorway to the bed. The door to the bathroom was slightly ajar. Light shone through the crack and fell on Jerry's shirt and pants, folded neatly on the floor.

She rushed to the door. "Jerry, are you in there?"

He was, and she did not scream. She took in the horrible sight as if it were what she had expected. He was so pale, his head hanging limp and unnatural. His ravaged arms, cut upward from the wrists, lay flat on the edges of the tub, palms downward and spread. The water, dark with his blood, was still warm on her hand where it dropped to his breast.

"I'll take you outside," Harper said.

"I can stay," she said. His razor lay on the edge of the tub.

Harper knelt down and reached for one of the mutilated wrists to feel for a pulse. She pulled his hand away and replaced it with her own. But she was not feeling for life, which she knew was gone. She was trying for a last and hopeless communion with the man she knew and lost and now somehow knew again.

His note lay half in a puddle of water and blood under the sink. She saw it first and picked it up when Harper went into the other room to call for help. The words ran on the page. The letters, though, were strong and straight, the way they had learned to make them in the schoolhouses on the plains.

"I read somewhere," it said, "that the words of a dying man are usually believed. Please believe this. I never wanted to mean any harm."

FOUR
After Action

33

Janet's business on the Coast stretched on and on. By the time she came home the weather had finally broken. The wind blew in, brisk and maritime. The low, dense sky had given way. August had become October in one clean break. The flags, which had hung damp and motionless in the stifling calm, now snapped sharply in the breeze. The monument's shadow was as distinct as truth and falsehood.

He had told her nothing of the Birch incident. It pleased him this time to be able to cope without her. Fran was out of his life, and he recorded this fact with the security section as soon as he was sure. She was just one more secret to be kept by the man she left because his secrets were not enough to sustain her. He had no regrets.

The evening Janet finally arrived, he was looking through his photographs. He had found them in the attic, still in the original box in which he had shipped them from Saigon. To his surprise, the heat and neglect had not yellowed them very much. The crowded street scenes, the cyclo drivers and waifs, the panorama of red tile roofs from the window of Janet's flat. He had laid out three from the collection on the table. One was of a little girl at a well-stocked black-market stall at the outdoor market. Her smiling, alien face was surrounded by tubes of Colgate, cartons of Kools, and bottles of whiskey. Another showed Harper and Bartlow, the station chief, lounging in tennis whites at the *Cercle Sportif,* Harper's right hand held up with two fingers raised in what had become the ambivalent sign of victory or peace. The last photograph was cruel and ugly, the broken bodies of three little children laid out in a village square. An American GI stood distractedly over them, a rifle in one hand angled upward from his hip, a transistor radio braced against his ear in the other.

"I thought you might have been thinking about me," Janet said when she saw them there. "Our wedding pictures perhaps. The silly look on your face."

He carried her bags in from the front steps, mixed her a drink, and brought it to her as she washed up.

"I did miss you," he said, setting the glass down on the Formica sink where she was rinsing the glaze of travel from her face. "Missed you terribly."

"It looks as if you got on well enough," she said. "I expected a shambles."

"I held together," he said, "barely."

"It feels good to be back."

"I guess you had a great success."

"I guess," she said. "What's wrong, Richard? I promise I won't be off on these trips often. I'll be doing most of my work right here from now on."

"The young man I worked with in Tokyo," he said, "he killed himself."

"Oh, Richard," she said, turning to him. "That's why you had those pictures. You are too hard on yourself, too punishing." She took him by the hand and led him back into the living room, where she put the Beethoven quartets on the player and turned the volume low.

"We had seen him in contact with the Russians," Harper said, "confronted him with it. Apparently it was more than he could handle."

"I should have been with you."

"It was important for me to get through it alone," he said. "I proved something about myself."

"How terrible for you, though. And you didn't even say. I knew something was wrong sometimes when we talked on the phone. You aren't very good at hiding your feelings, you know."

"I've thought of leaving the Agency."

"What would you do? Have you had an offer?"

"That isn't the important part of it," he said. "There are things I could do. Business. Teaching, perhaps. The Hill. They're always impressed by lapsed spooks. The question is whether I am fit for the work. They plan to move me into another section, pure analysis. It's a promotion. It takes me out of operations altogether, but I'll always know that side of it. Giving the assets numbers in the cables won't keep me from seeing their faces. I've been too close."

She slid away to show him the seriousness of her expression.

"You must stay there," she said. "You must stay simply because it isn't easy for you. Too many of the others, it doesn't bother them."

"And I suffer and atone," he said. But he was not really putting up an argument. She had said what he wanted her to say. She had named his true mission.

"You know what happens when the good say nothing," she said.

He reached over to the table and picked up a grotesque photograph, the one taken in the village he had identified as a staging area, the village the jets had bombed.

"I think I would like to have this framed," he said.

"As a reminder," she said.

"We can hang it somewhere just a little out of the way where the guests won't see it. But somewhere where I will see it from time to time. And you too."

"Promise me that if you ever have a crisis again and I am away you will call for me," she said. "I'll come to you no matter what. That will be my part."

"I promise," he said, easy in their conspiracy. And he kissed her. But Harper remembered the way Birch lay in the water, arms resting flat on porcelain, palms down, as if he held them steady there for the final fluttering. He remembered the gashes cut deep up the wrists from where the electrodes were always attached. He remembered the last telephone call when Birch whispered his signal of distress. And so Harper did not give a name to the message that would summon her. If sometime in the future they came together in his travail, if their common purpose brought them closer for a trying moment, he would not have a word for it. The word was corrupted. And they were pure.

Harper's part of the after-action review was somewhat limited. A committee was convened, and Harper had a seat on it. His specific assignment was only to write the damage assessment. The committee, though, was attentive to his more general interpretations, which he felt duty-bound to provide in meetings and memoranda. In the end, the committee accepted his analysis to a remarkable extent.

On the question of the death itself, the forensic evidence seemed clear. It had been a suicide. All the windows of the motel room were locked from the inside. The front door had always been within the desk clerk's view. He testified that no one had come in or out. Harper himself was completely beyond suspicion; Miss Lutkin had logged Birch's telephone call and recorded it. Harper's time was accounted for. The razor was Birch's; he had brought it from home for a single purpose. Of course, any violent death can be subject to doubts. If the Hill or the conspiracy theorists had gotten hold of this one, they might have found ambiguities. But there had not been much interest; intelligence abuses had by then lost their romance.

The committee had to defer to Harper's conclusion, against his own interest, that no one in the Agency but himself had had access to the code word. On the other points, he held the group by force of argument.

He asserted that Birch, by predisposition enhanced by training, had lost the ability to distinguish between reality and his own self-protective imagination. He was able to frustrate the polygraph because, in a sense, he always believed what he said. Believed simply because he had said it, said it because he had to. But when Birch came under suspicion from our side, he crumbled. He was forced to recognize the duplicity of his own mind and was unable in the end to separate it out.

Kerzhentseff, the argument went, never knew that Birch had doubled him in Tokyo. The dismantling of Black Body was strong, though admittedly not conclusive, evidence of this. If one assumed Kerzhentseff's ignorance, Birch's state of mind, and Harper's assertion that he had not initiated the last series of contacts, then it followed that Kerzhentseff or one of his minions had made the key call to Birch. This made some sense. Since the Black Body operation, the Soviets' operations had been in disarray. Kerzhentseff was newly back in the States, and his mission was undoubtedly to set up a new network. Birch was a likely candidate, a proven item.

Birch, in the meantime, was restive with ordinary service. He had fantasies of getting back into the action. That much was indicated by the Agency's files and by Birch's own word. So when the call came, he simply convinced himself that he was repeating his Tokyo experience. The code word rose in his imagination as the justification for undertaking the new mission. In a sense, he did hear it, but it was spoken by his own inner voice.

Harper's approach had the virtue of both clearing his own name and of ascribing all malice to the Russian. Harper used his analysis to lobby forcefully for Donna Birch, whose strength in the face of repeated revelations about her husband's secret life had impressed everyone concerned. To Harper's surprise, the representatives of the counterintelligence group were among his strongest allies. He had no delusion that this would in the future foreclose them from a reinterpretation. Counterintelligence never closes a file. But he was satisfied that for the moment he was able to put together a consensus that at once redeemed the memory of Jerry Birch and, in return for her signature on a release of Agency liability which no one felt she would press in any event, provided for his widow a modest patriot's pension.

True to the reasoning he had discussed with Janet, Harper made the point to the committee of review that Birch's death should serve as a lesson to the Agency as a whole. The regime of disinformation and radical doubt had a fatal potential. It could not be pushed upon the frail without consequences. In the future, he said, programs of disinformation must be undertaken in the knowledge that men are not born to lie, that intelligence itself is—and had been since the penetrating skepticism of Hume—based on certain unprovable assumptions that could, as well as any other designation, take the name of trust. Harper thought he was quite eloquent in this recitation. The others on the committee commended him for it. He was politic enough, however, not to insist that this conclusion be incorporated in the committee's final report. He said that he, for one, trusted that what could not for obvious reasons be written would nonetheless be taken to heart.

The damage assessment itself was a simpler matter. Black Body had been shut down, and Birch had been in a position to reveal to Kerzhentseff the ruse. But even if he had, this would not have been presently harmful because the technical side had discovered the solution and provided countermeasures. Slyly, Harper pointed out that at this juncture it would be helpful if Kerzhentseff did learn of the trick played upon him in Tokyo because then he might be tempted to put the now compromised system back into operation. Harper did not, of course, make any flat recommendation of another mission against the Russian. But his subtle suggestion did not go unnoticed.

As for other information Birch might have provided in his confused final days, Harper concluded that he was simply not privy to anything of much consequence: the names of certain personnel assigned to his security facility, which the Russians could have as easily obtained by a discreet surveillance of the gate; Harper's own identity, which was no longer closely held; the description of various polygraph techniques, which would be of limited importance since the equipment was available on the open market and the methods as universal as distrust.

In short, Harper wrote, not without a sadness that all who knew Harper's friendship with Birch could appreciate, the damage sustained in the whole misadventure was truly negligible.

34

I have heard it said that the moment when man established himself as a new species, unique among the animals, was the moment when he first used his primitive, guttural language to lie. It may have been for his amusement, but I prefer to think it was for survival. Until then he was bound; pure fact closed him in. Once he learned to dissemble, the world was his to create and destroy. And ever since that time, I suppose, there have been men such as Harper ready—for whatever reason— to play upon the nostalgia for a time of plain and bestial fact, and to call this a plea for human progress.

We let him have his way in the aftermath of the Birch affair. The truth of it has now been officially established, established as a state religion is: until the dogma cramps a king. The matter is closed. The files reflect no minority views, just as Harper wanted.

And Harper himself thrives. A testament to the flabby respectability that has come to dominate the Agency and the society it serves, he has been promoted to the top leadership group. His handling of the Birch investigation is considered to have been successful. No flap ensued. I have even heard it called "deft," of all things. So now Richard Harper is, in all but seniority, my equal. He is privy to the whole range of delicate information about our strengths and weaknesses, information that could destroy us. He is, in short, more dangerous than ever.

I consider this a personal failure of the first order. As long ago as the episode in Vietnam, I singled Harper out as a menace to our purposes. At the time, of course, it did not seem to be of any special consequence. Harper was then an operative of no particular stature. I called for his termination as a matter of principle. His response to the killing of Tri and the others had demonstrated that he was incapable of coping with doubt. He insisted on placing the blame for the deaths. He demanded satisfaction, a closed file, established truth.

Our institutional reaction was typically ambivalent. We rejected his demands for investigation and vindication, but we kept him on the payroll and gave him a second chance. Moreover, when Harper finally settled down and decided that the way to handle doubt was to deny its legitimacy, this was taken as a sign of rehabilitation. His view was said to be refreshing, useful. The time, of course, was ripe for such foolishness. His derision of systematic doubt became something of an intellectual fashion at Langley. Our sophisticated machines and technical means of observation had given us a false sense of precision.

When Harper achieved his great success in Tokyo, his career took off, and my concern about him became all the more pressing. Like Kerzhentseff, I recognized that in espionage we cannot afford to wait until jeopardy blossoms into harm. And if I needed any confirmation of the rightness of my assessment, it came when I became acquainted with Harper's theory of convergence. Even more than the business about the Mocking God, this revealed the depth of the man's flaw. It is a seditious idea. It obscures the fundamental difference between the United States and her enemies. Our enemies do not share our values and never will. They care nothing for humanism or the rule of law; they will use our beliefs against us if we allow them to. The evil they intend to do us is immutable, and we must be resolute against them. Dreams of mutuality are fatal.

The man who does not see this can ultimately be broken by their will.

It is not necessary to believe that Harper has already turned traitor to recognize the pitiful inadequacy of his analysis of the Birch affair, though it would be folly to rule this explanation out. (One need only ask oneself who reaped advantage in Tokyo and again at home from these strange episodes to see the outlines of a conspiracy to push Harper into a position where he could do us greatest harm.) Note the way he leaped to the assumption that his own operation had in no way been compromised. The Soviets, by his hypothesis, made the contact with Birch that led to the sighting at the Hay-Adams. But they did not speak the decisive word. This, Harper wants us to believe, Birch supplied himself. The conspiracy went no deeper than the darkness of one man's secret selves. How convenient. Had Birch been held to have known he was working for the Soviets and not the Americans, this would have cast doubt on Harper's judgment of the man, even upon the meaning of their great Tokyo success. So all blame goes to Birch in the defenselessness of his grave. I let Harper have his way in the committee of inquiry, but not because I believed this nonsense. Unless you have the superiority of force to destroy an adversary, it is folly to provoke him with a wound.

Harper's failure in the Birch affair went further. He refused to push to the crisis point the other obvious explanation of the known facts, for to have done so would have required him to throw himself against individuals in the Agency, as implacable as Kerzhentseff, whose strength Harper deathly feared.

First the matter of motive: Why would anyone in the Agency have wanted to put Birch into play? To fail to find a motive is never more than a failure of imagination. Human nature is profligate with malign purposes. Something as straightforward as professional envy and ambition, perhaps. Harper was—and is—on the rise; his career stands in others' way. But on a different level, Harper might have realized that there are those in the Agency like myself who, while they have no further personal ambitions, see his way of thinking as a threat to what we hold dear. It has been no secret that some of us adamantly object to his naïve vision of the world and his crabbed and cowardly sense of constraint. Harper should have known, even without my unsuccessful effort to taunt and pressure him on this point, that the whole Birch affair might have been arranged to break him. It might have aimed to force him to extreme measures such as contacting Kerzhentseff directly or trying to subvert Birch's story in order to protect himself. He was ready for a misstep, and it would have destroyed him. Harper was in a corner that, but for Birch's lapse of memory and untimely death, he could not have escaped without revealing some of his dangerous weaknesses.

Now as to means: It would have been quite easy for certain of us in the Agency to create the situation that put Harper to the test. By his own account, Birch spoke the code word on the telephone once in Tokyo. It was the night when Birch, coming apart, called Harper to that crazed, drunken meeting in the park. True, Harper never saw the words in the logs of the taps on Birch's phone. But he was too quick to assume from this that it had been inaudible to the machines and men monitoring. What if the word had been overheard? The soundman sitting on the tap might have withheld it at first to protect Harper's secret means of retaining complete control of the operation. Birch might have mentioned it to the Agency psychiatrists who interviewed him upon his return from Tokyo. Miss Lutkin (who had access to the logs and could have altered them to eliminate the word and cover her discovery) might have felt an obligation to report the mysterious reference. She is an admirably loyal woman who knows this quality is precious and rare. And whenever she has suspicions she always does precisely what she did in this instance. Miss Lutkin brings her doubts and evidence directly to me.

Brigaded with this potent bit of information and aware of Harper's fortuitous assignment to an indoctrination evening with the Sisters surveilling Kerzhentseff, I moved quickly. It was hardly a challenge for me to interest the Soviets in Birch again. I have for years been in touch with the network of competition and exchange. I know these people, how to reach them. I know them as intimately as I know my own soul.

Harper should have realized how simple it was to lead Kerzhentseff to clear the dead drop in the Hay-Adams alley. And with the code word, my simple call to Birch put that half of the strategy in motion.

Harper, of course, did not face up to these possibilities. It was another example of what makes him so unsuited for the position he occupies. He derides what he calls the Cult of the Mocking God, but this is only his excuse for failing to meet a threat head-on. He is a frightened little man, and as such he is ready to make a separate peace with any evil that promises to leave him unharmed.

The next time, perhaps, chance will not favor Harper so. The next time my instrumentality might not fail, as Birch did, at the crucial moment. I am confident that ultimately Harper will be broken. Even after I am gone, there will always be men as clever as Kerzhentseff ready to exploit him and, I hope and trust, men of my iron resolve ready to intervene. And when Harper finally falls and he hears the mocking laughter, it will come from no god at all. The laughter will be mine.